The Genetics and Development of Scoliosis

Kenro Kusumi · Sally L. Dunwoodie
Editors

The Genetics and
Development of Scoliosis

 Springer

Editors
Kenro Kusumi
School of Life Sciences
Arizona State University
P.O. Box 874501
Tempe AZ 85287-4501
USA

Sally L. Dunwoodie
Developmental Biology Division
Victor Chang Cardiac Research Institute
University of New South Wales
Lowy Packer Building
405 Liverpool St.
Darlinghurst NSW 2010
Australia

ISBN 978-1-4419-1405-7   e-ISBN 978-1-4419-1406-4
DOI 10.1007/978-1-4419-1406-4
Springer New York Dordrecht Heidelberg London

Library of Congress Control Number: 2009941779

Printed on acid-free paper

Springer is part of Springer Science+Business Media (www.springer.com)

The editors thank the patients and their families who have participated in scoliosis genetic studies and the research collaborators who have made these efforts possible

Preface

Scoliosis is a lateral curvature of the spine that is frequently encountered by health-care professionals. Scoliosis has historically been categorized into congenital, neuromuscular, and idiopathic forms, and related curves include kyphosis, kypho-scoliosis, and lordosis. Patients affected by scoliosis are concerned about prognosis, associated health conditions, and recurrence risks. Developmental genetic studies of the spine and linkage and family-based association studies have led to recent advances in understanding the genetic etiology of idiopathic and congenital scoliosis. Advances in genotyping and sequencing technology promise to further increase our understanding of the heterogeneous group of disorders involving spinal curvatures.

The inspiration for this volume was derived from the invited session, *Straightening Out the Curves: Understanding the Genetics Basis of Idiopathic and Congenital Scoliosis* organized at the 2008 American College of Medical Genetics, Annual Clinical Genetics Meeting in Phoenix, AZ, USA. The goals of that session were to bring leading researchers of both congenital and idiopathic scoliosis to present the current state of research and to compare potential shared developmental and genetic mechanisms. Following up on the discussions from this session, this volume presents the recent advances in studies of early spinal development and how disruptions in embryonic segmentation can lead to congenital vertebral defects. This volume also describes a recently defined clinical classification system for congenital vertebral disorders, based on identification of mutations in genes regulating segmentation. In addition, recent reports of genetic loci predisposing patients to develop juvenile and adolescent idiopathic scoliosis are presented, and key clinical features are reviewed. Finally, there is discussion of how genetic heterogeneity and gene–environment interactions may contribute to congenital scoliosis and isolated vertebral malformations.

Our understanding of the genetic and developmental mechanisms underlying scoliosis is rapidly evolving, and our goal in editing *The Genetics and Development of Scoliosis* was to provide researchers, clinicians, and students with the emerging views in this field.

Tempe, Arizona   *Kenro Kusumi*
Darlinghurst, New South Wales   *Sally L. Dunwoodie*

Contents

Contributors

Peter G. Alexander, Ph.D. Department of Orthopaedic Surgery, Center for Cellular and Molecular Engineering, University of Pittsburgh School of Medicine, Pittsburgh, PA 15219, USA, pea9@pitt.edu

Benjamin Alman, M.D. Hospital for Sick Children and University of Toronto, Toronto, Ontario, Canada benjamin.alman@sickkids.ca

Alberto Santiago Cornier, M.D., Ph.D. Department of Molecular Medicine, Hospital de la Concepción, San Germán, Puerto Rico; Ponce School of Medicine, Ponce, Puerto Rico, scornier@hospitalconcepcion.org

Walter Eckalbar School of Life Sciences, Arizona State University, Tempe, AZ 85287, USA walter.eckalbar@asu.edu

Rebecca E. Fisher, Ph.D. Department of Basic Medical Sciences, The University of Arizona College of Medicine–Phoenix in partnership with Arizona State University, Phoenix, AZ 85004, USA; School of Life Sciences, Arizona State University, Tempe, AZ 85287, USA, rfisher@email.arizona.edu

Philip F. Giampietro, M.D., Ph.D. Department of Pediatrics, Waisman Center, The University of Wisconsin-Madison, Madison, WI 53705-9345, USA, pfgiampietro@pediatrics.wisc.edu

Kenro Kusumi, Ph.D. School of Life Sciences, Arizona State University, Tempe, AZ 85287, USA; Department of Basic Medical Sciences, The University of Arizona College of Medicine–Phoenix in partnership with Arizona State University, Phoenix, AZ 85004, USA, kenro.kusumi@asu.edu

Nancy H. Miller, M.D. Department of Orthopaedic Surgery, Musculoskeletal Research Center, The Children's Hospital and University of Colorado Denver, Aurora, CO 80045, USA, nancy.hadley-miller@ucdenver.edu

Olivier Pourquié, Ph.D. Département de Biologie Cellulaire et Développement, Institut de Génétique et de Biologie Moléculaire et Cellulaire (IGBMC), Illkirch, F-67400 France; Inserm, U964, Illkirch, F-67400 France; CNRS, UMR7104, Illkirch, F-67400 France; Université de Strasbourg, Strasbourg, F-67000 France, pourquie@igbmc.fr

Alan Rawls, Ph.D. School of Life Sciences, Arizona State University, Tempe, AZ 85287, USA; Department of Basic Medical Sciences, The University of Arizona College of Medicine–Phoenix in partnership with Arizona State University, Phoenix, AZ 85004, USA, alan.rawls@asu.edu

Swarkar Sharma, Ph.D. Texas Scottish Rite Hospital for Children, Dallas, TX 75219, USA, swarkar.sharma@tsrh.org

Rocky S. Tuan, Ph.D. Department of Orthopaedic Surgery, Center for Cellular and Molecular Engineering, University of Pittsburgh School of Medicine, Pittsburgh, PA 15219, USA, rst13@pitt.edu

Peter D. Turnpenny, M.B.Ch.B. Clinical Genetics Department, Royal Devon & Exeter Hospital and Peninsula Medical School, Exeter EX1 2ED, UK, peter.turnpenny@rdeft.nhs.uk

Carol A. Wise, Ph.D. Texas Scottish Rite Hospital for Children, Dallas, TX 75219, USA, carol.wise@tsrh.org

Chapter 1
Genetic Regulation of Somite and Early Spinal Patterning

Kenro Kusumi, Walter Eckalbar, and Olivier Pourquié

Introduction

The spine forms the core of the vertebrate body plan and consists of complex repeating segments of musculoskeletal tissues, i.e., bone, cartilage, skeletal muscle, ligaments, and tendons, together with vasculature, lymphatics the spinal cord, and peripheral nerves. Key to this development is the formation of transient embryonic structures called somites, which will be discussed further below. As will be detailed in upcoming chapters, the spine plays a central biomechanical role, and disruptions in the structure of the vertebral column can have profound clinical consequences. The development of the spine takes place through a series of embryological events, detailed below and shown in Fig. 1.1.

(1) Formation of paraxial mesoderm
(2) Segmental pre-patterning in the presomitic mesoderm

    a. Notch pathway genes
    b. Wnt pathway genes
    c. FGF signaling
    d. Interactions between Notch, Wnt, and FGF pathway genes
    e. Prepatterning at the wavefront or determination front
    f. FGF and Wnt signaling and the determination front
    g. Retinoic acid and the determination front

K. Kusumi (✉)
School of Life Sciences, Arizona State University, Tempe, AZ 85287, USA;
Department of Basic Medical Sciences, The University of Arizona College of Medicine–Phoenix in Partnership with Arizona State University, Phoenix, AZ 85004, USA
e-mail: kenro.kusumi@asu.edu

O. Pourquié (✉)
IGBMC (Institut de Génétique et de Biologie Moléculaire et Cellulaire), Département de Biologie Cellulaire et Développement, Illkirch, F-67400 France; Inserm, U964, Illkirch, F-67400 France; CNRS, UMR7104, Illkirch, F-67400 France; Université de Strasbourg, Strasbourg, F-67000 France
e-mail: pourquie@igbmc.fr

K. Kusumi, S.L. Dunwoodie (eds.), *The Genetics and Development of Scoliosis*,
DOI 10.1007/978-1-4419-1406-4_1, © Springer Science+Business Media, LLC 2010

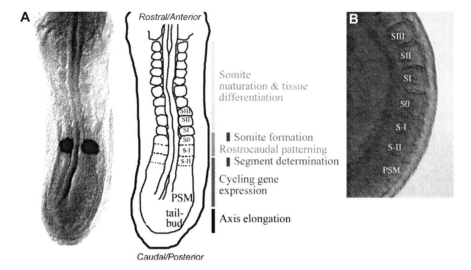

Fig. 1.1 Regions of developmental activity during somitogenesis. (**a**) Diagram illustrating an early somite-stage mouse embryo at 8.5 dpc (equivalent to a 22-day human embryo) with *Mesp2* expression visualized by whole-mount *in situ* hybridization. Regions of specific developmental activity are shown on the *right* and include (i) axis elongation, (ii) cycling gene expression, (iii) segment determination, (iv) rostrocaudal specification, (v) somite formation, (vi) somite maturation and tissue differentiation. (**b**) Lateral image of a 9.5-dpc mouse embryo. PSM, presomitic mesoderm; S0, newly forming somite; SI, newly formed somite I, S–I, region of PSM to next form a somite. Embryos are oriented rostrally toward the *top*

(3) Rostral–caudal orientation and somite segment formation
(4) Determination of myogenic and chondrogenic precursors within somites
(5) Division of somites into four compartments: the myotome, which generates embryonic skeletal muscle, the dermotome, which gives rise to the dermis, the syndetome, which gives rise to tendons, and the sclerotome, which contains somitic cells that give rise to the axial skeleton
(6) Resegmentation of the sclerotome, a migratory cell population that reshapes into vertebrae and ribs
(7) Embryonic skeletal muscle formation from the myotomal blocks
(8) Patterning of the peripheral nervous system and vasculature based on the somite template

In this chapter, we will focus our discussion on the first two steps, and the later steps will be described in the following chapter.

Formation of Paraxial Mesoderm

The paraxial mesoderm is formed in two phases. In the first phase, gastrulation, paraxial mesodermal cells are formed when cells of the epiblast ingress through the primitive streak in the caudal distal tip of the embryo. These epiblast cells undergo a transition from epithelial to mesenchymal organization, as regulated by genes in the Wnt, FGF, and BMP pathways, and disruption of these genes lead to failure of mesodermal formation (reviewed in Arnold and Robertson 2009, Ciruna and Rossant 2001). In the mouse model system, gastrulation generates the paraxial mesoderm up to the 30–31 somite stage at approximately 10 days *post coitum* (dpc; Wilson and Beddington 1996).

In the second phase, the paraxial mesoderm is formed from the structure called the tailbud. During the last stages of gastrulation, the primitive streak regresses and eventually disappears when the neural tube closes, leaving a region of mesenchymal cells in the tailbud adjacent to the ventral ectodermal ridge (VER, Tam and Tan 1992). The VER is required for continued generation of the paraxial mesoderm, and ablation of this tissue leads to premature truncation of the axis (Goldman et al. 2000). The VER appears to be the source of signaling molecules that regulate BMPs, which are required for continued proliferation of mesenchymal cells to contribute to presomitic mesoderm. Disruptions of the morphogenetic processes in the tailbud may underlie the defects observed in caudal regression syndrome or caudal agenesis (OMIM 600145). In this disorder, there is premature truncation of the axial skeleton, particularly in the thoraco-lumbar region, suggesting a defect in either the generation of paraxial mesoderm from the tailbud or the transition from the first to second phase of mesodermal formation. Currently, the genetic regulation of the transition from gastrulation to tailbud formation of paraxial mesoderm is not well understood.

Segmental Prepatterning in the Presomitic Mesoderm

The unsegmented paraxial mesoderm at the caudal end of the embryo is referred to as the presomitic mesoderm (PSM). At the rostral end of the PSM, segments of paraxial mesoderm are budded off to form the epithelial somites. This process will be discussed in greater detail in Chapter 2. While the unsegmented paraxial mesoderm does not reveal any morphologically obvious processes, prepatterning of the somite segments and their rostral–caudal compartmentalization are being determined by complex genetic regulatory networks within the PSM. Describing the current state of our understanding of this prepatterning and the implications for scoliosis and vertebral anomalies is the central aim of this chapter.

Before any genes were associated with somitogenesis, this iterative process was a favorite focus of mathematical modeling. The process of somite formation involves the conversion of a steady stream of tissue in the PSM into discrete pulses or somite segments. Mathematical models that describe this process usually rely on an "escapement" or device that stores energy and releases it in a series of regular, discrete periodic blocks or segments. One model that has become a very useful framework for understanding modern genetic findings is the "clock and wavefront" escapement model (Cooke and Zeeman 1976; Cooke 1981). In this model, cells within the PSM are sequentially recruited into successive somite formation due to an abrupt transition of their properties taking place at the level of a "wavefront," which travels caudally as somitogenesis proceeds. The periodicity of the transition that leads to somite formation is controlled by a "clock" or an oscillator that interacts with the wavefront to shift cells from one state to another, introducing the discontinuous borders required for segmentation. Modern additions to this model propose that the oscillator controls the periodic response of the PSM cells to external signals causing the transition between the two stable cell states (Goldbeter et al. 2007).

The first evidence that an oscillator may actually exist was the identification of periodic expression of the transcription factor *c-hairy1* in the chick PSM (Palmeirim et al. 1997). A dynamic wave of *c-hairy1* mRNA expression was observed to sweep from the caudal to the rostral end of the PSM concurrent with the formation of each somite. Based on this observation of cycling *c-hairy1* expression, the molecular realization of the oscillator mechanism proposed in the clock and wavefront model was termed the "segmentation clock" (Fig. 1.2). Research by a number of investigators has identified genes in the mammalian segmentation clock, including human cycling genes, as described further below and in Table 1.1.

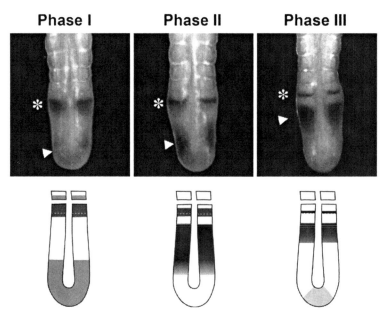

Fig. 1.2 Cycling gene expression in the presomitic mesoderm. Cycling gene expression is shown by Phases I–III, as defined previously (Pourquié and Tam 2001). *Top panels* show the cycling gene expression of the *Nrarp* gene in 10.5-dpc mouse embryos, as visualized by whole-mount *in situ* hybridization (Sewell et al. 2009). *Asterisks* mark the rostral region of *Nrarp* expression in S0, the nascent somite, which condenses from Phase I to III. White *arrowheads* mark the caudal region of *Nrarp* expression in the unsegmented presomitic mesoderm, which progresses rostrally during each somite cycle. Diagrams below highlight key features of each phase of *Nrarp* expression. Embryos are oriented rostrally toward the *top*

Notch Pathway Genes

The first cycling genes to be identified in mammals were members of the Notch pathway, including the modulator and glycosyltransferase *lunatic fringe* (*Lfng*) (Forsberg et al. 1998, McGrew et al. 1998, Aulehla and Johnson 1999) and the

Table 1.1 Cycling and oscillatory genes in mammalian somitogenesis

Gene symbol	Description	Signaling pathway(s)	Reference(s)
Cycling genes in somitogenesis			
Axin2	axin2	Wnt	Aulehla et al. (2003)
Dact1	Dapper homolog 1, antagonist of beta-catenin	Wnt	Dequéant et al. (2006) and Suriben et al. (2006)
Dkk1	Dickkopf homolog 1	Wnt	Dequéant et al. (2006)
Dusp4	Dual specificity phosphatase 4	EGF, FGF	Chu et al. (1996), Dequéant et al. (2006) and Niwa et al. (2007)
Dusp6	Dual specificity phosphatase 6	FGF, MAPK/ERK	Klock and Herrmann (2002), Dequéant et al. (2006) and Niwa et al. (2007)
Hes1	Hairy and enhancer of split 1	Notch	Jouve et al. (2000)
Hes7	Hairy and enhancer of split 7	Notch	Bessho et al. (2001a)
Hey2	Hairy/enhancer-of-split related with YRPW motif 2	Notch	Leimeister et al. (2000a)
Heyl (Hey3)	Hairy/enhancer-of-split related with YRPW motif-like	Notch	Leimeister et al. (2000b) and Kusumi et al. (2004)
Hspg2	Perlecan (heparan sulfate proteoglycan 2)	FGF	Sharma et al. (1998) and Dequéant et al. (2006)
Lfng	Lunatic fringe, *O*-fucosylpeptide 3-beta-*N*-acetylglucosaminyltransferase	Notch	McGrew et al. (1998), Aulehla and Johnson, (1999)
Myc	Myelocytomatosis oncogene	ErbB, Jak-STAT, MAPK, TGF-beta, Wnt	Feng et al. (2002) and Sansom et al. (2007)
Nkd1	Naked cuticle 1 homolog	Notch, Wnt	Ishikawa et al. (2004)
Nrarp	Notch-regulated ankyrin repeat protein	Notch and Wnt	Dequéant et al. (2006) and Sewell et al. (2009)
Ptpn11 (Shp2)	Protein tyrosine phosphatase, non-receptor type 11	FGF, IL-1, -2, and -6	Saxton and Pawson (1999) and Dequéant et al. (2006)
Sp5	Trans-acting transcription factor 5	T-box, Wnt	Harrison et al. (2000), Weidinger et al. (2005) and Dequéant et al. (2006)
Spry2	Sprouty homologue 2	FGF, Jak-STAT	Shim et al. (2005) and Dequánt et al. (2006)
Spry4	Sprouty homologue 4	FGF, Jak-STAT	Hayashi et al. (2009)
Tnfrsf19	Tumor necrosis factor receptor superfamily, member 19	JNK, TNF, Wnt	Eby et al. (2000) and Dequéant et al. (2006)
Other oscillatory genes in somitogenesis			
Maml3	Mastermind-like 3	Notch	William et al. (2007)
Nkd2	Naked cuticle 2 homolog	Wnt	William et al. (2007) and Zhang et al. (2007)

basic helix–loop–helix effectors *Hes1*, *Hes7*, *Hey5*, and *Hey1* (Palmeirim et al. 1997, Leimeister et al. 2000a, Bessho et al. 2001a). Genetic evidence from targeted mutants also highlighted the requirement of Notch signaling in segmentation, including *Notch1* (Conlon et al. 1995), recombination signal binding protein for immunoglobulin kappa J region/mammalian Suppressor of Hairless (*Rbpj*, Oka et al. 1995; Barrantes et al. 1999), *delta-like 1* (*Dll1*, Hrabe de Angelis et al. 1997; Barrantes et al. 1999), *Lfng* (Evrard et al. 1997, Zhang et al. 2007, Shifley et al. 2008), *delta-like 3* (*Dll3*, Kusumi et al. 1998, Dunwoodie et al. 2002, Kusumi et al. 2004, Sewell et al. 2009), *presenilin 1* (*Psen1*, Shen et al. 1997), *Hes7* (Bessho et al. 2001b), and *peptide-O-fucosyltransferase 1* (*Pofut1*, Schuster-Gossler et al. 2009). Interestingly, targeted mutation of the mouse orthologue of *c-hairy1*, *Hes1*, does not display any defects in somitogenesis (Ishibashi et al. 1995). However, mutations in both cycling genes such as *Hes7* and noncycling genes such as *Dll1* and *Dll3* produce segmental defects. This suggests that the segmentation clock has the capacity to compensate for loss of single cycling genes and that noncycling genes are also required to maintain the segmentation clock.

Activation of the Notch1 receptor leads to cleavage of the Notch intracellular domain (Notch-ICD) that translocates to the nucleus. Oscillatory changes in Notch-ICD protein levels have been observed by immunohistochemistry, indicating cyclical activation of the Notch signaling pathway (Huppert et al. 2005, Morimoto et al. 2005). POFUT1 activates Notch by adding O-linked fucose, which are later modified by LFNG, to serine or threonine residues. *Lfng* expression is activated by Notch1, and in turn, LFNG inhibits the activation of *Notch1* signaling, establishing a negative feedback loop (Dale et al. 2003). In *Lfng* null mutants, the lack of negative feedbacks leads to constitutive Notch-ICD expression within the PSM (Morimoto et al. 2005). However, when *Lfng* is constitutively expressed in the PSM, cycling expression of endogenous *Lfng* and *Hes7* is still observed (Serth et al. 2003). This suggests that cycling inhibition of Notch signaling by LFNG is required for the segmentation clock but that ectopic inhibition of Notch signaling by LFNG is not sufficient to shut down the network.

The HES protein family establishes key negative feedback loops in the Notch component of the segmentation clock (Bessho et al. 2003). The expression levels of the transcriptional repressors *Hes7*, *Hes1*, and *Hey2* oscillate in the mouse PSM (Jouve et al. 2000, Bessho et al. 2001a). However, *Hes1* and *Hey2* null mutants have no somitic phenotype (Ishibashi et al. 1995; Ohtsuka et al. 1999, Gessler et al. 2002). Only *Hes7* null mutants display defects in somitogenesis and expression of cycling genes such as *Lfng*, and upregulation of transcription from *Hes7* promoter (Bessho et al. 2001a). This supports the hypothesis that *Hes7* represses its own transcription, as well as the transcription of *Lfng* in the PSM (Bessho et al. 2003). The *Lfng* promoter contains a CSL/*Rbpj* binding site, which is required for regulation by Notch signaling, and binding sites for bHLH proteins (Morales et al. 2002, Cole et al. 2002). Oscillations of *Lfng* in the caudal PSM are lost when these regulatory regions are deleted, which leads to ubiquitous activation of Notch in the PSM and severe vertebral anomalies in the cervico-thoraco-lumbar region but not the sacral and caudal region, suggesting differential requirement of *Lfng* oscillation along the rostral–caudal axis (Shifley et al. 2008).

Due to the increasing complexity of the genetic network regulating somitogenesis, mathematical models are playing an increasing role in our understanding of the segmentation clock in the mouse embryo. Most models have been based on delayed negative feedback loops established through HES-family members (Hirata et al. 2004, Lewis 2003, Monk 2003). One key variable in these models is the half-life of the *Hes* RNA and protein. In mouse, HES7 and HES1 proteins are degraded by a proteasome-dependent mechanism and protein half-life is estimated to be approximately 22 min (Bessho et al. 2003, Hirata et al. 2002, Hirata et al. 2004). The time-delay mathematical model predicts that increasing the half-life of the HES7 protein would disrupt cycling, and when the half-life of the HES7 protein was genetically modified to increase from 22 to 30 min, this prediction was validated (Hirata et al. 2004).

Wnt Pathway Genes

Several lines of evidence have demonstrated a regulatory role for Wnt signaling in somitogenesis. First, a negative feedback inhibitor of the Wnt pathway, *Axin2*, was observed to display cycling expression in mouse PSM (Aulehla et al. 2003). Second, the *Wnt3a* vestigial tail (*vt*) hypomorphic mutant displays disruptions in the cycling expression of the Notch gene *Lfng* and *Axin2*, suggesting that Wnt signaling may be required for signaling of two pathways (Aulehla et al. 2003). Third, microarray-based screens for genes with oscillatory expression have identified a number of Wnt pathway genes with the periodicity of the segmentation clock in the mouse PSM (Dequéant et al. 2006) or in human mesenchymal stem cells (William et al. 2007). Genes identified by Dequéant et al. (2006) clustered into different groups, with one cluster primarily made up of genes linked to the Wnt signaling pathway. This group includes the trans-acting transcription factor 5 (*Sp5*), myelocytomatosis oncogene (*Myc*), and the negative feedback inhibitors *Axin2* and *Dkk1* (Dequéant et al. 2006, Wedinger et al. 2005, He et al. 1998, Glinka et al. 1998).

In canonical Wnt signaling, receptor activation leads to the stabilization of beta-catenin, which then enters the nucleus to activate transcription of its target genes. Oscillatory expression of Wnt inhibitors, such as *Dkk1* or *Dact1* (Dequéant et al. 2006, Suriben et al. 2006), would then be expected to lead to oscillatory expression of beta-catenin levels. Mutations in beta-catenin, like those in the *Wnt3a* ligand, lead to disruption of gene oscillations in both the Notch and the Wnt pathways (Dunty et al. 2008). However, when *beta-catenin* is constitutively expressed, gene oscillations are not affected for either the Wnt or the Notch pathways (Dunty et al. 2008, Aulehla et al. 2008). Thus, oscillatory *beta-catenin* expression is necessary but not sufficient to drive the segmentation clock.

Further evidence that Wnt signaling is necessary but not necessarily the master regulator of the segmentation clock comes from targeted mutations. *Axin2* null mutants, like those of *Hes1*, do not display any somite phenotype (Yu et al. 2005). However, disruptions of the Wnt inhibitor *Dkk1* and the Wnt target genes *Sp5* and *Myc* cause segmental defects (MacDonald et al. 2004, Harrison et al. 2000,

Trumpp et al. 2001, Camenisch et al. 2000). In chick embryos treated with CKI-7, an inhibitor of the LRP binding factor casein kinase 1, the period of somitogenesis increases from one somite pair per 90 min to one somite pair per 115–120 min (Gibb et al. 2009). Thus, both Wnt and Notch signaling are necessary for the segmentation clock, but due to the complexity of the mechanism, the hierarchy of gene interactions still needs further investigation.

Fibroblast Growth Factor (FGF) Signaling

Recent studies have identified that members of the FGF pathway oscillate in phase with the Notch pathway cyclic genes (reviewed in Dequéant and Pourquié 2008). These cycling genes include the antagonists of FGF signaling, sprouty homolog 2 (*Spry2*), and dual specificity phosphatase 6 (*Dusp6*). Targets of the FGF pathway in the mouse, including snail homolog 1 (*Snai1*) and snail homolog 2 (*Snai2*), and *Dusp4*, a negative feedback inhibitor of the FGF pathway, also display an oscillating expression pattern (Dale et al. 2006, Niwa et al. 2007). Oscillations of another negative inhibitor of the FGF pathway *Sprouty4*, which oscillates in phase with the Notch pathway gene *Lfng*, are lost in *Hes7*-null mutants, suggesting a possible link between the FGF and the Notch pathways (Hayashi et al. 2009).

Interactions Between Notch, Wnt, and FGF Pathway Genes

Not all genes fall only into one pathway or another, and genes that interact with many pathways are particularly interesting within the segmentation clock. Among these genes is the Notch-regulated ankyrin repeat protein (*Nrarp*), which is a target of Notch and acts as a negative regulator of the Notch pathway (Lamar et al. 2001, Krebs et al. 2001, Pirot et al. 2004). *Nrarp* also functions to stabilize the Wnt pathway transcription factor, member lymphoid enhancer binding factor 1 (*Lef1*), thus positively regulating the Wnt pathway (Ishitani et al. 2005). *Nrarp* displays a pattern of cycling expression distinct from *Axin2* and *Lfng* patterns, which is disrupted in *Dll3*, *Lfng*, and *Wnt3a* null embryos (Sewell et al. 2009). Naked cuticle 1 homolog (*Nkd1*), a previously identified cyclic gene which acts as an inhibitor of Wnt signaling, is also regulated by Notch signaling (Ishikawa et al. 2004). However, targeted mutations in *Nkd1* and *Nkd2* do not produce any defects in somitogenesis (Zhang et al. 2007). Genes that serve as possible links between the Notch and the Wnt pathways are of great interest in terms of understanding the regulation of the segmentation clock.

While the cycling patterns of Notch- and FGF-associated genes appear in phase in the PSM, this co-expression does not necessarily indicate co-regulation. For example, cycling expression of *Spry2* is retained in the Notch pathway *Rbpj* null mutants (Dequéant et al. 2006). Furthermore, cycling expression of the FGF *Dusp4* is observed in *Lfng, Dll1*, and *Rbpj* null mutants, where Notch signaling is inhibited in explanted mouse PSM with DAPT (Niwa et al. 2007). However, cycling expression of selected genes in the FGF, Notch, and Wnt pathways is disrupted with the conditional deletion of *fibroblast growth factor receptor 1* (*Fgfr1*) gene in the PSM (Niwa et al. 2007, Wahl et al. 2007). Cyclic expression of *Axin2* and *Spry2* is quickly lost in presomitic mesoderm explant cultures treated with the FGF inhibitor SU-5402. However, *Lfng* oscillations are lost only after a one-cycle delay in SU-5402-treated tails, and this delay suggests that Notch oscillations are indirectly regulated by FGF signaling (Wahl et al. 2007). By transitive logic, this suggests that FGF may act upstream of Wnt signaling, since Wnt signaling is required for *Lfng* oscillations (Aulehla and Herrmann 2004; Nakaya et al. 2005).

However, other lines of evidence do not support this hierarchy. Oscillations in the expression of the Notch target *Hes7* also require FGF signaling in mouse (Niwa et al. 2007). Cyclic expression of *Hes7* appears to be initiated by FGF signaling in the tailbud, while *Hes7* oscillations are maintained by Notch signaling in the more rostral PSM (Niwa et al. 2007). This two-step model proposed to control *Hes7* oscillations is consistent with the delay observed in *Lfng* oscillations when *Fgfr1* is being inhibited. However, in conditional *Fgfr1* mutants where expression of beta-catenin is constitutively stabilized, *Lfng* cycling expression is still observed (Aulehla et al. 2008). Notch oscillations are restored by constitutive beta-catenin in the absence of FGF signaling, which contradicts the idea that FGF signaling controls Wnt oscillations. Taken together, these experiments suggest that neither Wnt nor FGF signals control the oscillating expression of cyclic genes, such as *Lfng*. This, combined with the observations that oscillations in *Axin2* expression are maintained in Notch pathway mutants, suggests that none of these signaling pathways acts as the pacemaker for the segmentation clock (Aulehla et al. 2003).

It is possible that these three signaling pathways are controlled by a still unidentified upstream pacemaker. A more parsimonious explanation would be that the cycling genes in the segmentation clock belong to more than one signaling pathway and we have yet to understand how this complex network is organized. Just as mathematical models (Cooke and Zeeman 1976) were helpful in understanding the basic mechanisms regulating somitogenesis, modern systems biology approaches may help us to identify the key regulatory interactions between the genes we have identified.

Prepatterning at the Wavefront or Determination Front

The clock and wavefront model (Cooke and Zeeman 1976) postulated that interaction between the PSM wavefront and the oscillator recruited cells ready to form somites. This model placed the wavefront at the most rostral region of the PSM where the nascent somites are forming. However, heat-shock experiments in the frog challenged this localization and suggested that interaction between the wavefront and the oscillator occurs caudal to where it was envisioned (Elsdale et al.

1976). This wavefront was further refined through experiments in the chick, where sections of PSM tissue were rotated 180° and allowed to develop (Dubrulle et al. 2001). The boundary, rostral to which tissues failed to reorient and were fixed in rostro-caudal polarity, was called the determination front and defined the location at which PSM cells acquire the segmental identity. The location of the determination front coincides with the caudal boundary of mesoderm posterior 2 (*Mesp2*) expression, marking segmental prepatterning in the PSM. This will be discussed further in the following chapter.

The determination front in the PSM also corresponds with a morphological transition at the cellular level. The cells in the caudal PSM are a loose mesenchyme, while the cells rostral to the determination front are progressively epithelialized. Cell movements in the PSM also slow down once passing this transition (Delfini et al. 2005). This mesenchymal–epithelial transition and segmental patterning of the rostral PSM also correlate with a downregulation of Snail genes. The Snail genes are regulated by FGF signaling and in many systems they are associated with the mesenchymal state (Dale et al. 2006). At the determination front the downregulation of Snail genes is associated with the expression of adhesion molecules, such as integrins or cadherins, which progressively increase in abundance as cells in the rostral PSM become polarized (Linask et al. 1998, Horikawa et al. 1999). During this transition the basal lamina, made up of laminin and fibronectin, is deposited to surround the rostral PSM (Duband et al. 1987). The bHLH transcription factor paraxis, which is expressed rostral to the determination front, is also required for the mesenchymal–epithelial transition (Burgess et al. 1995). Paraxis functions to mediate the mesenchymal–epithelial transition by regulating the transcription of Rho GTPases, such as RAS-related C3 botulinum substrate 1 (*Rac1*) and cell division cycle 42 homolog (*Cdc42*) (Nakaya et al. 2004).

FGF and Wnt Signaling and the Determination Front

Experiments have shown that levels of both FGF and Wnt signaling appear to define the location of the determination front (Aulehla et al. 2003, Dubrulle et al. 2001, Sawada et al. 2001). An FGF gradient was observed for *Fgf8* mRNA levels in the PSM of chick, fish, and mouse embryos, with the highest levels at the caudal end (Dubrulle et al. 2001, Sawada et al. 2001, Dubrulle and Pourquié 2004a). The FGF8 protein gradient establishes an MAPK/AKT gradient along the PSM (Delfini et al. 2005, Sawada et al. 2001, Dubrulle and Pourquié 2004). Experiments disrupting the concentrations of FGF along the rostro–caudal PSM gradient in chicken embryos caused markers of the caudal PSM, such as *T*/Brachyury, to extend rostrally and caused markers of segmentation and differentiation, such as *paraxis, Mesp2*, or myogenic differentiation (*Myod1*), to be downregulated (Dubrulle et al. 2001, Delfini et al. 2005).

Redundancy in the FGF pathway has made experiments on *Fgf8* loss-of-function mutants difficult to interpret. *Fgf8* is likely not the only ligand involved in the FGF-signaling gradient because a conditional deletion of *Fgf8* in mice does not display a segmentation phenotype (Perantoni et al. 2005). In addition to *Fgf8*, three other components of the FGF pathway have so far been identified that are expressed in the PSM and tailbud of the mouse: *Fgf3*, *Fgf4*, and *Fgf18* (Wahl et al. 2007). Treatment of chick embryos with pharmacological inhibitors of the FGF pathway causes the expression domain of genes expressed in the caudal PSM, such as *Fgf8*, to be shifted further caudally. Mouse mutants with a conditional deletion of *Fgfr1*, the sole FGF receptor identified so far in the PSM, in the paraxial mesoderm also results in a similar caudal shift of the expression domains of *Fgf8* and mesogenin 1 (*Msgn1*) (Wahl et al. 2007). Taken together, these experiments show that the FGF pathway likely functions to maintain the caudal identity of PSM cells (Dubrulle et al. 2001).

The Wnt ligand *Wnt3a* is also expressed in the tailbud and caudal PSM (Aulehla et al. 2003). As with FGF signaling, a gradient of beta-catenin has highest levels in the tailbud and extends toward the determination front (Aulehla et al. 2008). The expression of Wnt targets, such as *Axin2*, also identifies the caudal to rostral Wnt-signaling gradient (Aulehla et al. 2003). It has been shown that expression of FGF ligands is dependent on Wnt signaling from the observation that *Fgf8* expression is lost in *Wnt3a* mutants. In the PSM, the beta-catenin gradient affects the position of the determination front and defines the size of the oscillatory field. Downregulation of Wnt signaling at the determination front is likely required for segmentation to proceed normally, as suggested by beta-catenin gain-of-function mutants in the PSM that inhibit activation of *Mesp2* targets (Dunty et al. 2008, Aulehla et al. 2008).

Transcription of *Fgf8* mRNA in PSM precursors in the tailbud terminates when their descendents enter the caudal PSM. *Fgf8* mRNA progressively decays as these cells progress into the more rostral regions of the PSM. This process establishes an *Fgf8* mRNA gradient that is converted into a gradient of FGF signaling activity (Delfini et al. 2005, Sawada et al. 2001, Dubrulle and Pourquié 2004). The Wnt gradient is thought to be regulated by a similar mechanism (Aulehla and Herrmann 2004). As the FGF and the Wnt mRNA and proteins decay in the PSM, the determination front is displaced caudally. The rate of somitogenesis is then set by the speed of this displacement. This mechanism maintains the coupling between axis elongation and segmentation in the developing embryo.

Despite FGF and Wnt pathway activation following a caudal-to-rostral gradient, paradoxically the expression of some of their targets genes displays cycling expression. The beta-catenin gain-of-function mutants have demonstrated that nuclear beta-catenin plays a role in controlling the maturation of PSM cells. However, these experiments have also shown that oscillations of Wnt targets, such as *Axin2*, are the result of oscillating inputs independent of canonical beta-catenin and the FGF pathways (Aulehla et al. 2008).

Retinoic Acid and the Determination Front

The location of the determination front is also regulated by a rostral-to-caudal gradient of retinoic acid (RA), opposite to that of the Wnt and FGF gradients

(Vermot and Pourquié, 2005, Sirbu and Duester, 2006, Diez del Corral et al. 2003, Moreno and Kintner 2004). The RA gradient is established by limiting the expression of its biosynthetic enzyme source, *Raldh2*, to the rostral most PSM and somitic region (Niederreither et al. 2002a, b). Through RA response element (RARE)-LacZ reporter experiments in mouse, it was found that RA signaling was limited to the rostral PSM and somites, and absent from the caudal PSM and tailbud, where the enzyme *Cyp26*, which degrades RA, hence serving as a sink, is expressed in response to FGF signals and degrades RA (Vermot et al. 2005, Abu-Abed et al. 2001). Treatment of explants of the caudal PSM of chicken embryos with RA antagonists downregulates the expression of *Fgf8*. Also, *Raldh2* expression is downregulated when an FGF8-soaked bead is grafted to the PSM (Diez del Corral et al. 2003). Furthermore, the *Fgf8* domain extends to the rostral PSM in mouse *Raldh2* mutants and in RA-deprived chicken or quail embryos (Vermot et al. 2005, Diez del Corral et al. 2003). The opposing gradients of FGF and RA are found in *Xenopus* and thus appear to be conserved across vertebrates (Moreno and Kintner 2004).

The mutual inhibition of the opposing FGF and RA gradients was proposed to localize the determination front (Fig. 1.3, Diez del Corral et al. 2003). Unfortunately, RA signaling does not appear to be necessary for segmentation, as somites are still formed in *Raldh2*-mutant mice lacking RA signaling (Niederreither et al. 2002a, b). Furthermore, a caudal shift in the RARE-lacZ domain is not observed in conditional mutations of *Fgfr1*. This suggests the RA gradient is being opposed

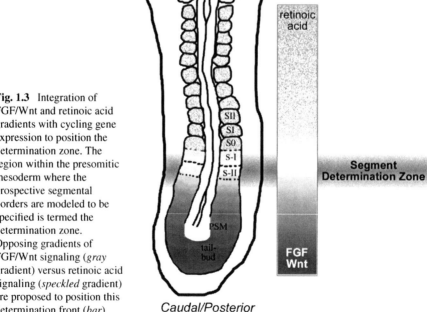

Fig. 1.3 Integration of FGF/Wnt and retinoic acid gradients with cycling gene expression to position the determination zone. The region within the presomitic mesoderm where the prospective segmental borders are modeled to be specified is termed the determination zone. Opposing gradients of FGF/Wnt signaling (*gray* gradient) versus retinoic acid signaling (*speckled* gradient) are proposed to position this determination front (*bar*)

by more than the FGF gradient (Wahl et al. 2007). It remains to be determined if the Wnt gradient alone can antagonize RA signaling.

Evidence also points to a role for RA controlling the symmetry of somitogenesis (Vermot and Pourquié, 2005, Sirbu and Duester, 2006). In embryos lacking RA, asymmetry between the spacing of somites on the left and right sides of the embryo can develop. If allowed to develop, these types of lateralized defects could result in spinal curvatures, due to unilateral vertebral malformations. Studies have demonstrated that RA inhibits the ability of the paraxial mesoderm to respond to asymmetric signaling downstream of the left-side determinant Nodal, which could disrupt the symmetry of somitogenesis (Vermot and Pourquié 2005).

Comparison with Somitogenesis in Humans

The periodicity of somite formation and cycling gene expression is species-specific, with average periods ranging from 30 min in zebrafish to 90 min in the chick and 2 h in the mouse (Saga & Takeda, 2001). The rate at which somites are produced in human embryos can be estimated from the Carnegie collection, giving an average of 4–4.5 somites each day between embryonic days 22 and 28 (Carnegie stages 10–13, calculated from O'Rahilly et al. 1979). This would give an estimated periodicity of cycling genes of 5.3–6 h. Molecular analysis of cycling genes in human synchronized cell cultures has identified a periodicity of approximately 5–6 h for *HES1* and other genes, which is consistent with these estimates (William et al. 2007). The increased cycling period in human compared to other mammals could have arisen by a number of mechanisms, including the following:

(1) The half-life of auto-regulatory genes is longer in humans.
(2) The threshold levels for activation or repression may differ between species.
(3) The latency period or time required for transcriptional activation of cycling genes may be longer in humans.

The effects of these quantitative changes are particularly suitable for systems biological analysis of the complex genetic network.

The slower pace of human somitogenesis will need to be considered in studying the mechanisms that lead to vertebral birth defects. If it takes over twice as long for a somite to form in humans compared to other mammals such as the mouse, the window of susceptibility to environmental insults is greater in humans. Toxicological studies have identified teratogenic factors, including hypoxia and carbon monoxide exposure. Hypoxia causes skeletal defects including hemivertebrae, vertebral fusions, fragmented vertebral bodies, and rib malformations in the mouse model (Ingalls and Curley 1957; Murakami and Kameyama 1963; Rivard et al. 1979). Maternal carbon monoxide exposure results in congenital spinal deformities directly related to dose and time of exposure (Loder et al. 2000, Murray et al. 1979, Schwetz et al. 1979, Singh et al. 1984).

Carbon monoxide is a commonly encountered occupational and environmental pollutants, with rates of smoking reported among pregnant women at 11% (National Center for Health Statistics 2009). However, since somitogenesis occurs early in gestation starting at embryonic day 20 when many women may be unaware of the pregnancy, it is also relevant that greater percentage of women in the general population (18%) are smokers (National Center for Health Statistics 2009).

With the identification of somitogenesis genes that are disrupted in human spondylocostal dysostosis and spondylothoracic dysplasia (*DLL3, HES7, LFNG, MESP2;* Bulman et al. 2000, Sparrow et al. 2008, Sparrow et al. 2006, Whittock et al. 2004, Cornier et al. 2008), we are at the cusp of being able to study gene–environment interactions. Understanding the genetic pathways regulating somitogenesis will help us to identify the quantitative effects of environmental insults such as hypoxia or carbon monoxide on gene expression and cell function.

Acknowledgments We would like to thank Mary-Lee Dequéant and Will Sewell for their contributions in preparing this chapter.

References

Abu-Abed, S., Dollé, P., Metzger, D., Beckett, B., Chambon, P., and Petkovich, M. 2001. The retinoic acid-metabolizing enzyme, CYP26A1, is essential for normal hindbrain patterning, vertebral identity, and development of posterior structures. Genes Dev. 15:226–240.

Arnold, S.J. and Robertson, E.J. 2009. Making a commitment: cell lineage allocation and axis patterning in the early mouse embryo. Nat. Rev. Mol. Cell. Biol. 10:91–103.

Aulehla, A. and Johnson, R.L. 1999. Dynamic expression of lunatic fringe suggests a link between notch signaling and an autonomous cellular oscillator driving somite segmentation. Dev. Biol. 207:49–61.

Aulehla, A., Wehrle, C., Brand-Saberi, B., Kemler, R., Gossler, A., Kanzler, B., and Herrmann, B.G. 2003. Wnt3a plays a major role in the segmentation clock controlling somitogenesis. Dev. Cell 4:395–406.

Aulehla, A. and Herrmann, B.G. 2004. Segmentation in vertebrates: clock and gradient finally joined. Genes Dev. 18:2060–2067.

Aulehla, A., Wiegraebe, W., Baubet, V., Wahl, M.B., Deng, C., Taketo, M., Lewandoski, M., and Pourquié, O. 2008. A beta-catenin gradient links the clock and wavefront systems in mouse embryo segmentation. Nat. Cell Biol. 10:186–193.

Barrantes, I., Elia, A.J., Wünsch, K., Hrabe de Angelis, M.H., Mak, T.W., Rossant, J., Conlon, R.A., Gossler, A., and de la Pompa, J.L. 1999. Interaction between notch signalling and lunatic fringe during somite boundary formation in the mouse. Curr. Biol. 9:470–480.

Bessho, Y., Miyoshi, G., Sakata, R., and Kageyama, R. 2001a. Hes7: a bHLH-type repressor gene regulated by Notch and expressed in the presomitic mesoderm. Genes Cells 6:175–185.

Bessho, Y., Hirata, H., Masamizu, Y., and Kageyama, R. 2001b. Dynamic expression and essential functions of Hes7 in somite segmentation. Genes Dev. 15:2642–2647.

Bessho, Y., Hirata, H., Masamizu, Y., and Kageyama, R. 2003. Periodic repression by the bHLH factor Hes7 is an essential mechanism for the somite segmentation clock. Genes Dev. 17: 1451–1456.

Bulman, M.P., Kusumi, K., Frayling, T.M., McKeown, C., Garrett, C., Lander, E.S., Krumlauf, R., Hattersley, A.T., Ellard, S., and Turnpenny, P.D. 2000. Mutations in the human delta homologue, DLL3, cause axial skeletal defects in spondylocostal dysostosis. Nat. Genet. 24:438–441.

Burgess, R., Cserjesi, P., Ligon, K.L., and Olson, E.N. 1995. Paraxis: a basic helix-loop-helix protein expressed in paraxial mesoderm and developing somites. Dev. Biol. 168:296–306.

Camenisch, T.D., Spicer, A.P., Brehm-Gibson, T., Biesterfeldt, J., Augustine, M.L., Calabro, A. Jr., Kubalak, S., Klewer, S.E., and McDonald, J.A. 2000. Disruption of hyaluronan synthase-2 abrogates normal cardiac morphogenesis and hyaluronan-mediated transformation of epithelium to mesenchyme. J. Clin. Invest. 106:349–360.

Chu, Y., Solski, P.A., Khosravi-Far, R., Der, C.J., and Kelly, K. 1996. The mitogen-activated protein kinase phosphatases PAC1, MKP-1, and MKP-2 have unique substrate specificities and reduced activity in vivo toward the ERK2 sevenmaker mutation. J. Biol. Chem. 271(11):6497–6501.

Ciruna, B. and Rossant, J. 2001. FGF signaling regulates mesoderm cell fate specification and morphogenetic movement at the primitive streak. Dev. Cell 1(1):37–49.

Cole, S.E., Levorse, J.M., Tilghman, S.M., and Vogt, T.F. 2002. Clock regulatory elements control cyclic expression of lunatic fringe during somitogenesis. Dev. Cell 3:75–84.

Conlon, R.A., Reaume, A.G., and Rossant, J. 1995. Notch1 is required for the coordinate segmentation of somites. Development. 121:1533–1545.

Cooke, J. and Zeeman, E.C. 1976. A clock and wavefront model for control of the number of repeated structures during animal morphogenesis. J. Theor. Biol. 58:455–476.

Cooke, J. 1981. The problem of periodic patterns in embryos. Phil. Trans. R. Soc. Lond. B 295:509–524.

Cornier, A.S., Staehling-Hampton, K., Delventhal, K.M., Saga, Y., Caubet, J.F., Sasaki, N., Ellard, S., Young, E., Ramirez, N., Carlo, S.E., Torres, J., Emans, J.B., Turnpenny, P.D., and Pourquié, O. 2008. Mutations in the MESP2 gene cause spondylothoracic dysostosis/Jarcho-Levin syndrome. Am. J. Hum. Genet. 82(6):1334–1341.

Dale, J.K., Maroto, M., Dequéant, M.L., Malapert, P., McGrew, M., and Pourquié, O. 2003. Periodic notch inhibition by lunatic fringe underlies the chick segmentation clock. Nature 421:275–278.

Dale, J.K., Malapert, P., Chal, J., Vilhais-Neto, G., Maroto, M., Johnson, T., Jayasinghe, S., Trainor, P., Herrmann, B., and Pourquié. O. 2006. Oscillations of the snail genes in the presomitic mesoderm coordinate segmental patterning and morphogenesis in vertebrate somitogenesis. Dev. Cell 10:355–366.

Delfini, M.C., Dubrulle, J., Malapert, P., Chal, J., and Pourquie, O. 2005. Control of the segmentation process by graded MAPK/ERK activation in the chick embryo. Proc. Natl. Acad. Sci. U.S.A. 102:11343–11348.

Dequéant, M.L., Glynn, E., Gaudenz, K., Wahl, M., Chen, J., Mushegian, A., and Pourquié, O. 2006. A complex oscillating network of signaling genes underlies the mouse segmentation clock. Science 314:1595–1598.

Dequéant ML, and Pourquié O. 2008. Segmental patterning of the vertebrate embryonic axis. Nat. Rev. Genet. 9(5):370–382.

Diez del Corral, R., Olivera-Martinez, I., Goriely, A., Gale, E., Maden, M.,, and Storey, K. 2003. Opposing FGF and retinoid pathways control ventral neural pattern, neuronal differentiation, and segmentation during body axis extension. Neuron 40:65–79.

Duband, J.L., Dufour, S., Hatta, K., Takeichi, M., Edelman, G.M., and Thiery, J.P. 1987. Adhesion molecules during somitogenesis in the avian embryo. J. Cell Biol. 104:1361–1374.

Dubrulle, J., McGrew, M.J., and Pourquie, O. 2001. FGF signaling controls somite boundary position and regulates segmentation clock control of spatiotemporal Hox gene activation. Cell 106:219–232.

Dubrulle, J. and Pourquie, O. 2004a. fgf8 mRNA decay establishes a gradient that couples axial elongation to patterning in the vertebrate embryo. Nature 427:419–422.

Dubrulle, J. and Pourquie, O. 2004b. Coupling segmentation to axis formation. Development 131:5783–5793.

Dunty, W.C., Jr., Biris, K.K., Chalamalasetty, R.B., Taketo, M.M., Lewandoski, M., and Yamaguchi, T.P. 2008. Wnt3a/beta-catenin signaling controls posterior body development by coordinating mesoderm formation and segmentation. Development 135:85–94.

Dunwoodie, S.L., Clements, M., Sparrow, D.B., Sa, X., Conlon, R.A., and Beddington, R.S. 2002. Axial skeletal defects caused by mutation in the spondylocostal dysplasia/pudgy gene Dll3 are associated with disruption of the segmentation clock within the presomitic mesoderm. Development 129:1795–1806.

Eby, M.T., Jasmin, A., Kumar, A., Sharma, K., and Chaudhary, P.M. 2000. TAJ, a novel member of the tumor necrosis factor receptor family, activates the c-Jun N-terminal kinase pathway and mediates caspase-independent cell death. J. Biol. Chem. 275:15336–15342.

Elsdale, T., Pearson, M., and Whitehead, M. 1976. Abnormalities in somite segmentation following heat shock to Xenopus embryos. J. Embryol. Exp. Morphol. 35:625–635.

Evrard, Y.A., Lun, Y., Aulehla, A., Gan, L., and Johnson, R.L. 1998. Lunatic fringe is an essential mediator of somite segmentation and patterning. Nature 394:377–381.

Feng, X.-H., Liang, Y.-Y., Liang, M., Zhai, W., and Lin, X. 2002. Direct interaction of c-Myc with Smad2 and Smad3 to inhibit TGF-beta-mediated induction of the CDK inhibitor p15(Ink4B). Molec. Cell 9:133–143.

Forsberg, H., Crozet, F., and Brown, N.A. 1998. Waves of mouse Lunatic fringe expression, in four-hour cycles at two-hour intervals, precede somite boundary formation. Curr. Biol. 8: 1027–1030.

Gessler, M., Knobeloch, K.P., Helisch, A., Amann, K., Schumacher, N., Rohde, E., Fischer, A., and Leimeister, C. 2002. Mouse gridlock: no aortic coarctation or deficiency, but fatal cardiac defects in Hey2 -/- mice. Curr. Biol. 12(18):1601–1604.

Gibb, S., Zagorska, A., Melton, K., Tenin, G., Vacca, I., Trainor, P., Maroto, M., and Dale, J.K. 2009 Interfering with Wnt signaling alters the periodicity of the segmentation clock. Dev. Biol. 330:21–31

Glinka, A., Wu, W., Delius, H., Monaghan, A.P., Blumenstock, C., and Niehrs, C. 1998. Dickkopf-1 is a member of a new family of secreted proteins and functions in head induction. Nature 391:357–362.

Goldbeter, A., Gonze, D., and Pourquié, O. 2007. Sharp developmental thresholds defined through bistability by antagonistic gradients of retinoic acid and FGF signaling. Dev. Dyn. 236: 1495–1508.

Goldman, D.C., Martin, G.R., and Tam, P.P. 2000. Fate and function of the ventral ectodermal ridge during mouse tail development. Development 127:2113–2123.

Harrison, S.M., Houzelstein, D., Dunwoodie, S.L., and Beddington, R.S. 2000. Sp5, a new member of the Sp1 family, is dynamically expressed during development and genetically interacts with Brachyury. Dev. Biol. 227:358–372.

Hayashi, S., Shimoda, T., Nakajima, M., Tsukada, Y., Sakumura, Y., Dale, J.K., Maroto, M., Kohno, K., Matsui, T., and Bessho, Y. 2009. Sprouty4, an FGF inhibitor, displays cyclic gene expression under the control of the notch segmentation clock in the mouse PSM. PLOS. 4(5):e5603.

He, T.C., Sparks, A.B., Rago, C., Hermeking, H., Zawel, L., da Costa, L.T., Morin, P.J., Vogelstein, B., and Kinzler, K.W. 1998. Identification of c-MYC as a target of the APC pathway. Science 281:1509–1512.

Hirata, H., Yoshiura, S., Ohtsuka, T., Bessho, Y., Harada, T., Yoshikawa, K., and Kageyama, R. 2002. Oscillatory expression of the bHLH factor Hes1 regulated by a negative feedback loop. Science 298:840–843.

Hirata, H., Bessho, Y., Kokubu, H., Masamizu, Y., Yamada, S., Lewis, J., and Kageyama, R. 2004. Instability of Hes7 protein is crucial for the somite segmentation clock. Nat. Genet. 36: 750–754.

Horikawa K, Radice G, Takeichi M, and Chisaka O. 1999. Adhesive subdivisions intrinsic to the epithelial somites. Dev. Biol. 215:182–189.

Hrabe de Angelis, M., McIntyre, J., 2nd, and Gossler, A. 1997. Maintenance of somite borders in mice requires the Delta homologue Dll1. Nature 386:717–721.

Huppert, S.S., Ilagan, M.X., De Strooper, B., and Kopan, R. 2005. Analysis of Notch function in presomitic mesoderm suggests a gamma-secretase-independent role for presenilins in somite differentiation. Dev. Cell 8:677–688.

Ingalls, T.H. and Curley, F.J. 1957. Principles governing the genesis of congenital malformations induced in mice by hypoxia. N. Engl. J. Med. 257:1121–1127.

Ishibashi, M., Ang, S.L., Shiota, K., Nakanishi, S., Kageyama, R., and Guillemot, F. 1995. Targeted disruption of mammalian hairy and Enhancer of split homolog-1 (HES-1) leads to up-regulation of neural helix-loop-helix factors, premature neurogenesis, and severe neural tube defects. Genes Dev. 9:3136–3148.

Ishikawa, A., Kitajima, S., Takahashi, Y., Kokubo, H., Kanno, J., Inoue, T., and Saga Y. 2004. Mouse Nkd1, a Wnt antagonist, exhibits oscillatory gene expression in the PSM under the control of Notch signaling. Mech. Dev. 121:1443–1453.

Ishitani, T., Matsumoto, K., Chitnis, A.B., and Itoh, M. 2005. Nrarp functions to modulate neural-crest-cell differentiation by regulating LEF1 protein stability. Nat. Cell Biol. 7:1106–1112.

Jouve, C., Palmeirim, I., Henrique, D., Beckers, J., Gossler, A., Ish-Horowicz, D., and Pourquié, O. 2000. Notch signalling is required for cyclic expression of the hairy-like gene HES1 in the presomitic mesoderm. Development 127:1421–1429.

Klock, A. and Herrmann, B.G. 2002. Cloning and expression of the mouse dual-specificity mitogen-activated protein (MAP) kinase phosphatase Mkp3 during mouse embryogenesis. Mech. Dev. 116(1–2):243–247.

Krebs, L.T., Deftos, M.L., Bevan, M.J., and Gridley, T. 2001. The Nrarp gene encodes an ankyrin-repeat protein that is transcriptionally regulated by the notch signaling pathway. Dev. Biol. 238:110–119.

Kusumi, K., Sun, E.S., Kerrebrock, A.W., Bronson, R.T., Chi, D.C., Bulotsky, M.S., Spencer, J.B., Birren, B.W., Frankel, W.N., and Lander, E.S. 1998. The mouse pudgy mutation disrupts Delta homologue Dll3 and initiation of early somite boundaries. Nat. Genet. 19(3): 274–278.

Kusumi, K., Mimoto, M.S., Covello, K.L., Beddington, R.S., Krumlauf, R., and Dunwoodie, S.L. 2004. Dll3 pudgy mutation differentially disrupts dynamic expression of somite genes. Genesis 39(2):115–121.

Lamar, E., Deblandre, G., Wettstein, D., Gawantka, V., Pollet, N., Niehrs, C., and Kintner, C. 2001. Nrarp is a novel intracellular component of the Notch signaling pathway. Genes Dev. 15:1885–1899.

Leimeister, C., Dale, K., Fischer, A., Klamt, B., Hrabe de Angelis, M., Radtke, F., McGrew, M.J., Pourquié, O., and Gessler, M. 2000a. Oscillating expression of c-hey2 in the presomitic mesoderm suggests that the segmentation clock may use combinatorial signaling through multiple interacting bHLH factors. Dev. Biol. 227:91–103.

Leimeister, C., Schumacher, N., Steidl, C., and Gessler, M. 2000b. Analysis of HeyL expression in wild-type and Notch pathway mutant mouse embryos. Mech. Dev. 98(1–2):175–178.

Lewis, J. 2003. Autoinhibition with transcriptional delay: a simple mechanism for the zebrafish somitogenesis oscillator. Curr. Biol. 13(16):1398–1408.

Linask, K.K., Ludwig, C., Han, M.D., Liu, X., Radice, G.L., and Knudsen, K.A. 1998. N-cadherin/catenin-mediated morphoregulation of somite formation. Dev. Biol. 202:85–102.

Loder, R.T., Hernandez, M.J., Lerner, A.L., Winebrener, D.J., Goldstein, S.A., Hensinger, R.N., Liu, C.Y., and Schork, M.A. 2000. The induction of congenital spinal deformities in mice by maternal carbon monoxide exposure. J. Pediatr. Orthop. 20:662–666.

MacDonald, B.T., Adamska, M., and Meisler, M.H. 2004. Hypomorphic expression of Dkk1 in the doubleridge mouse: dose dependence and compensatory interactions with Lrp6. Development. 131:2543–2552.

McGrew, M.J., Dale, J.K., Fraboulet, S., and Pourquié, O. 1998. The lunatic fringe gene is a target of the molecular clock linked to somite segmentation in avian embryos. Curr. Biol. 8:979–982.

Monk, N.A.M. 2003. Oscillatory expression of Hes1, p53 and NF-kB driven by transcriptional time delays. Curr. Biol. 13:1409–1413.

Morales, A.V., Yasuda, Y., and Ish-Horowicz, D. 2002. Periodic lunatic fringe expression is controlled during segmentation by a cyclic transcriptional enhancer responsive to notch signaling. Dev. Cell 3:63–74.

Moreno, T.A. and Kintner, C. 2004. Regulation of segmental patterning by retinoic acid signaling during Xenopus somitogenesis. Dev. Cell 6:205–218.

Morimoto, M., Takahashi, Y., Endo, M., and Saga, Y. 2005. The Mesp2 transcription factor establishes segmental borders by suppressing Notch activity. Nature 435:354–359.

Murakami, U. and Kameyama, Y. 1963. Vertebral malformations in the mouse fetus caused by maternal hypoxia during early stages of pregnancy. J. Embryol. Exp. Morphol. 11: 107–118.

Murray, F.J., Schwetz, B.A., Crawford, A.A., Henck, J.W., Quast, J.F., and Staples, R.E. 1979. Embryotoxicity of inhaled sulfur dioxide and carbon monoxide in mice and rabbits. J. Environ. Sci. Health 13:233–250.

Nakaya, M.A., Biris, K., Tsukiyama, T., Jaime, S., Rawls, J.A., and Yamaguchi, T.P. 2005. Wnt3a links left-right determination with segmentation and anteroposterior axis elongation. Development 132:5425–5436.

Nakaya, Y., Kuroda, S., Katagiri, Y.T., Kaibuchi, K., and Takahashi, Y. 2004. Mesenchymal-epithelial transition during somitic segmentation is regulated by differential roles of Cdc42 and Rac1. Dev Cell 7:425–438.

National Center for Health Statistics. 2009. Health, United States, 2008 with Chartbook. Hyattsville, MD: U.S. Department of Health and Human Services.

Niederreither, K., Fraulob, V., Garnier, J.M., Chambon, P., and Dolle, P. 2002a. Differential expression of retinoic acid-synthesizing (RALDH) enzymes during fetal development and organ differentiation in the mouse. Mech. Dev. 110:165–171.

Niederreither, K., Abu-Abed, S., Schuhbaur, B., Petkovich, M., Chambon, P., and Dollé, P. 2002b. Genetic evidence that oxidative derivatives of retinoic acid are not involved in retinoid signaling during mouse development. Nat. Genet. 31:84–88.

Niwa, Y., Masamizu, Y., Liu, T., Nakayama, R., Deng, C.X., and Kageyama, R. 2007. The initiation and propagation of Hes7 oscillation are cooperatively regulated by Fgf and notch signaling in the somite segmentation clock. Dev. Cell 13:298–304.

Ohtsuka, T., Ishibashi, M., Gradwohl, G., Nakanishi, S., Guillemot, F., and Kageyama, R. 1999. Hes1 and Hes5 as notch effectors in mammalian neuronal differentiation. EMBO J. 18: 2196–2207.

Oka, C., Nakano, T., Wakeham, A., de la Pompa, J.L., Mori, C., Sakai, T., Okazaki, S., Kawaichi, M., Shiota, K., Mak, T.W., and Honjo, T. 1995. Disruption of the mouse RBP-J kappa gene results in early embryonic death. Development. 121(10):3291–3301.

O'Rahilly, R., Muller, F., and Meyer, D.B. 1980. The human vertebral column at the end of the embryonic period proper. 1. The column as a whole. J. Anat. 131(Pt 3):565–575.

Palmeirim, I., Henrique, D., Ish-Horowicz, D., and Pourquié, O. 1997. Avian hairy gene expression identifies a molecular clock linked to vertebrate segmentation and somitogenesis. Cell 91: 639–648.

Perantoni, A.O., Timofeeva, O., Naillat, F., Richman, C., Pajni-Underwood, S., Wilson, C., Vainio, S., Dove, L.F., and Lewandoski, M. 2005. Inactivation of FGF8 in early mesoderm reveals an essential role in kidney development. Development 132:3859–3871.

Pirot, P., van Grunsven, L.A., Marine, J.C., Huylebroeck, D., and Bellefroid, E.J. 2004. Direct regulation of the Nrarp gene promoter by the Notch signaling pathway. Biochem. Biophys. Res. Commun. 322:526–534.

Pourquié, O., and Tam, P.P. 2001. A nomenclature for prospective somites and phases of cyclic gene expression in the presomitic mesoderm. Dev. Cell 1:619–620.

Rivard, C.H., Narbaitz, R., and Uhthoff, H.K. 1979. Time of induction of congenital vertebral malformations in human and mouse embryo. Orthop. Rev. 8:135–139.

Saga, Y. and Takeda, H. 2001. The making of the somite: molecular events in vertebrate segmentation. Nat. Rev. Genet. 2:835–845.

Sansom, O.J., Meniel, V.S., Muncan, V., Phesse, T.J., Wilkins, J.A., Reed, K.R., Vass, J.K., Athineos, D., Clevers, H., and Clarke, A.R. 2007. Myc deletion rescues Apc deficiency in the small intestine. Nature 446:676–679.

Sawada, A., Shinya, M., Jiang, Y.J., Kawakami, A., Kuroiwa, A., and Takeda, H. 2001. Fgf/MAPK signalling is a crucial positional cue in somite boundary formation. Development. 128:4873–4880.

Saxton, T.M. and Pawson, T. 1999. Morphogenetic movements at gastrulation require the SH2 tyrosine phosphatase Shp2. Proc. Natl. Acad. Sci. U.S.A. 96(7): 3790–3795.

Schuster-Gossler, K., Harris, B., Johnson, K.R., Serth, J., and Gossler, A. 2009. Notch signalling in the paraxial mesoderm is most sensitive to reduced Pofut1 levels during early mouse development. BMC Dev. Biol. 9:6.

Schwetz, B.A., Smith, F.A., Leong, B.K.J., and Staples, R.E. 1979. Teratogenic potential of inhaled carbon monoxide in mice and rabbits. Teratology 19:385–392.

Serth, K., Schuster-Gossler, K., Cordes, R., and Gossler, A. 2003. Transcriptional oscillation of Lunatic fringe is essential for somitogenesis. Genes Dev. 17:912–925.

Sewell, W., Sparrow, D., Gonzalez, D.M., Smith, A., Eckalbar, W., Gibson, J., Dunwoodie, S.L., and Kusumi, K. 2009. Cyclical expression of the Notch/Wnt regulator *Nrarp* requires *Dll3* function in somitogenesis. Dev Biol. 329:400–409.

Sharma, B., Handler, M., Eichstetter, I., Whitelock, J.M., Nugent, M.A., and Iozzo, R.V. 1998. Antisense targeting of perlecan blocks tumor growth and angiogenesis in vivo. J. Clin. Invest. 102:1599–1608.

Shen, J., Bronson, R.T., Chen, D.F., Xia, W., Selkoe, D.J., and Tonegawa, S. 1997. Skeletal and CNS defects in Presenilin-1-deficient mice. Cell. 89:629–639.

Shifley, E.T., Vanhorn, K.M., Perez-Balaguer, A., Franklin, J.D., Weinstein, M., and Cole, S.E. 2008. Oscillatory lunatic fringe activity is crucial for segmentation of the anterior but not posterior skeleton. Development 135:899–908.

Shim, K., Minowada, G., Coling, D.E., and Martin, G.R. 2005. Sprouty2, a mouse deafness gene, regulates cell fate decisions in the auditory sensory epithelium by antagonizing FGF signaling. Dev. Cell 8:553–564.

Singh, J., Aggison, L., and Moore-Cheatum, L. 1984. Teratogenicity and developmental toxicity of carbon monoxide in protein deficient mice. Teratology 48:149–159.

Sirbu, I.O. and Duester, G. 2006. Retinoic-acid signaling in node ectoderm and posterior neural plate directs left-right patterning of somitic mesoderm. Nat. Cell Biol. 8:271–277.

Sparrow, D.B., Chapman, G., Wouters, M.A., Whittock, N.V., Ellard, S., Fatkin, D., Turnpenny, P.D., Kusumi, K., Sillence, D., and Dunwoodie, S.L. 2006. Mutation of the LUNATIC FRINGE gene in humans causes spondylocostal dysostosis with a severe vertebral phenotype. Am. J. Hum. Genet. 78:28–37.

Sparrow, D.B., Guillén-Navarro, E., Fatkin, D., and Dunwoodie, S.L. 2008. Mutation of hairy-and-enhancer-of-split-7 in humans causes spondylocostal dysostosis. Hum. Mol. Genet. 17(23):3761–3766.

Suriben, R., Fisher, D.A., and Cheyette, B.N. 2006. Dact1 presomitic mesoderm expression oscillates in phase with Axin2 in the somitogenesis clock of mice. Dev. Dyn. 235(11):3177–3183.

Tam, P.P. and Tan, S.S. 1992. The somatogenetic potential of cells in the primitive streak and the tail bud of the organogenesis-stage mouse embryo. Development 115:703–715.

Trumpp, A., Refaeli, Y., Oskarsson, T., Gasser, S., Murphy, M., Martin, G.R., and Bishop, J.M. 2001. c-Myc regulates mammalian body size by controlling cell number but not cell size. Nature 414:768–773.

Vermot, J. and Pourquie, O. 2005. Retinoic acid coordinates somitogenesis and left-right patterning in vertebrate embryos. Nature 435:215–220.

Wahl, M.B., Deng, C., Lewandoski, M., and Pourquie, O. 2007. FGF signaling acts upstream of the NOTCH and WNT signaling pathways to control segmentation clock oscillations in mouse somitogenesis. Development 134:4033–4041.

Weidinger, G., Thorpe, C.J., Wuennenberg-Stapleton, K., Ngai, J., and Moon, R.T. 2005. The Sp1-related transcription factors sp5 and sp5-like act downstream of Wnt/beta-catenin signaling in mesoderm and neuroectoderm patterning. Curr. Biol. 15:489–500.

Whittock, N.V., Sparrow, D.B., Wouters, M.A., Sillence, D., Ellard, S., Dunwoodie, S.L., and Turnpenny, P.D. 2004. Mutated MESP2 causes spondylocostal dysostosis in humans. Am. J. Hum. Genet. 74:1249–1254.

William, D.A., Saitta, B., Gibson, J.D., Traas, J., Markov, V., Gonzalez, D.M., Sewell, W., Anderson, D.M., Pratt, S.C., Rappaport, E.F., and Kusumi, K. 2007. Identification of oscillatory genes in somitogenesis from functional genomic analysis of a human mesenchymal stem cell model. Dev. Biol. 305:172–186.

Wilson, V. and Beddington, R.S. 1996. Cell fate and morphogenetic movement in the late mouse primitive streak. Mech. Dev. 55:79–89.

Yu, H.M., Jerchow, B., Sheu, T.J., Liu, B., Costantini, F., Puzas, J.E., Birchmeier, W., and Hsu, W. 2005. The role of Axin2 in calvarial morphogenesis and craniosynostosis. Development 132:1995–2005.

Zhang, N., and Gridley, T. 1998. Defects in somite formation in lunatic fringe-deficient mice. Nature. 394:374–377.

Zhang, S., Cagatay, T., Amanai, M., Zhang, M., Kline, J., Castrillon, D.H., Ashfaq, R., Oz, O.K., and Wharton, K.A. Jr. 2007. Viable mice with compound mutations in the Wnt/Dvl pathway antagonists nkd1 and nkd2. Mol. Cell. Biol. 27(12):4454–4464.

Chapter 2
Development and Functional Anatomy of the Spine

Alan Rawls and Rebecca E. Fisher

Introduction

The vertebral column is composed of alternating vertebrae and intervertebral (IV) discs supported by robust spinal ligaments and muscles. All of these elements, bony, cartilaginous, ligamentous, and muscular, are essential to the structural integrity of the spine. The spine serves three vital functions: protecting the spinal cord and spinal nerves, transmitting the weight of the body, and providing a flexible axis for movements of the head and the torso. The vertebral column is capable of extension, flexion, lateral flexion (side to side), and rotation. However, the degree to which the spine is capable of these movements varies by region. These regions, including the cervical, the thoracic, the lumbar, and the sacrococcygeal spine, form four curvatures (Fig. 2.1). The thoracic and the sacrococcygeal curvatures are established in fetal development, while the cervical and the thoracic curvatures develop during infancy. The cervical curvature arises in response to holding the head upright, while the lumbar curvature develops as an infant begins to sit upright and walk. Congenital defects and degenerative diseases can result in exaggerated, abnormal curvatures. The most common of these include a thoracic kyphosis (or hunchback deformity), a lumbar lordosis (or swayback deformity), and scoliosis. Scoliosis involves a lateral curvature of greater than 10°, often accompanied by a rotational defect. To appreciate the potential underlying causes of scoliosis, we need to understand the cellular and genetic basis of vertebral column and skeletal muscle development from somites. In this chapter, we will review the embryonic development of the spine and associated muscles and link them to the functional anatomy of these structures in the adult.

A. Rawls and R.E. Fisher (✉)
School of Life Sciences, Arizona State University, Tempe, AZ 85287, USA;
Department of Basic Medical Sciences, The University of Arizona College of Medicine–Phoenix in Partnership with Arizona State University, Phoenix, AZ 85004, USA
e-mail: alan.rawls@asu.edu
e-mail: rfisher@email.arizona.com

K. Kusumi, S.L. Dunwoodie (eds.), *The Genetics and Development of Scoliosis*,
DOI 10.1007/978-1-4419-1406-4_2, © Springer Science+Business Media, LLC 2010

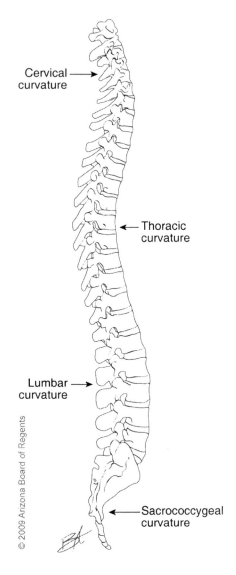

Fig. 2.1 Lateral view of the vertebral column, illustrating the spinal curvatures. Drawing by Brent Adrian

Cervical curvature

Thoracic curvature

Lumbar curvature

Sacrococcygeal curvature

Embryonic Origins of the Spine

The origins of the vertebral column, spinal musculature, and associated tendons are two rods of paraxial mesoderm that fill in the space on either side of the neural tube at the time of gastrulation. Beginning at 20 days *post coitus*, paraxial meso-derm undergoes segmentation in a rostral to caudal direction to form 42–44 pairs of somites, which can be subdivided into 4 occipital, 8 cervical, 12 thoracic, 5 lumbar, 5 sacral, and 8–10 coccygeal somites. The first occipital and the last 5–7 coccygeal somites disappear during embryonic development. Each somite will differentiate into four cell lineage-specific compartments that contribute to the verte-bral column and associated musculature: sclerotome (vertebrae and ribs), syndetome (tendons), myotome (skeletal muscle), and dermomyotome (dermis and skeletal muscle progenitor cells).

Somite formation can best be described as a continuous segmentation of mesenchymal cells from the rostral end of the paraxial mesoderm or the presomitic mesoderm (PSM) that lays down the embryonic cells that will give rise to the axial skeleton. Intrinsic to this process is (1) an oscillating clock controlling the timing of somitogenesis, (2) the formation of intersomitic boundaries, (3) mesenchymal to epithelial transition (MET), and (4) positional identity (e.g., rostral/caudal and dorsal/ventral). Experimental disruption in any one of the processes in vertebrate model organisms (e.g., mouse and chick) can lead to an axial skeletal dysmorphogenesis that is phenotypically consistent with scoliosis. The timing of somite formation and the determination of the site of boundary formation are established by the interactions between the Notch, Wnt, and FGF signaling pathways. This process is reviewed in Chapter 1. Here we will focus on the morphogenetic events associated with the physical separation of PSM during formation of the boundary, epithelialization, and positional identity.

Establishing the Intersomitic Boundary

Boundary formation occurs as somitic cells pull apart from the adjacent PSM. Depending on the animal, this varies from the simple cleavage of the PSM by fissures initiated along either the medial or the lateral surfaces as seen in *Xenopus* and zebrafish to a more dynamic ball-and-socket shape with a reshuffling of cells across the presumptive somite–PSM boundary in chicks (Wood and Thorogood 1994, Henry et al. 2000, Jiang et al. 2000, Kulesa and Fraser 2002, Afonin et al. 2006, Kulesa et al. 2007). The activity is an intrinsic property of the PSM, as it will occur in explants in the absence of the adjacent ectoderm and endoderm (Palmeirim et al. 1998). However, the underlying mechanism(s) remains poorly understood. In studies carried out in chick embryos, the fissure can be induced by activated Notch receptors and is stabilized by the presence of Lfng (Sato et al. 2002). Transcription factors *Mesp2* (and its chicken homologue *cMeso1*) and *Tbx18* have also been shown to play a role in forming boundaries (Saga et al. 1997, Buchberger et al. 1998, Tanaka and Tickle 2004, Takahashi and Sato 2008). Ectopic expression of either *cMeso1* or *Tbx18* is sufficient to induce ectopic fissures in chick PSM. Additional signals derived from the ventral PSM coordinate fissure formation in the dorsal PSM, though the nature of the signal remains poorly understood (Sato and Takahashi 2005). It is likely that the physical separation of cells at the fissure is related to differential changes in cell adhesion.

Somite Epithelialization

Cells of the newly formed somites undergo an increase in cell number, density, and expression of extracellular matrix proteins (reviewed in Tam and Trainor 1994,

Keynes and Stern 1988), resulting in the condensation of mesenchyme into an epithelial ball, surrounding a mesenchymal core, called the somitocoel. This occurs in a gradual process with the cells along the rostral edge of somite 0 becoming epithelia at the time of boundary formation (Dubrulle and Pourquié 2004). Epithelialization is complete with the formation of the next boundary (Fig. 2.2). The transcription factors paraxis and *Pax3* are required to direct MET in cells of somite +1 (Burgess et al. 1995, 1996, Schubert et al. 2001). Inactivation of paraxis results in somites formed of loose clusters of mesenchyme separated by distinct intersomitic boundary formation (Fig. 2.2). This reveals that MET is not required for boundary formation. However, the two events are temporally linked, suggesting that they are both responsive to the oscillating segmental clock. Candidate genes for linking the two are snail1 and 2 (*Snail* and *Snai2*), which are expressed in oscillating patterns in the PSM (Dale et al. 2006). Snail genes are transcriptional repressors that are able to block the transcription of paraxis and cell adhesion molecules associated with epithelialization (Batlle et al. 2000, Cano et al. 2000, Barrallo-Gimeno and Nieto 2005, Dale et al. 2006). Overexpression of snail2 will prevent cells from contributing to epithelium in somite +1. Thus switching off snail gene expression may be essential for the timing of MET.

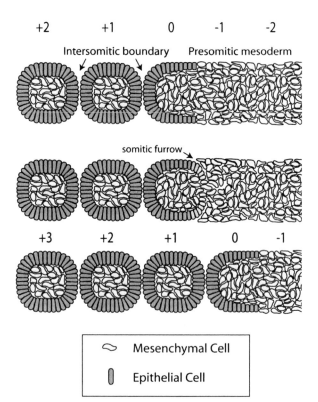

Fig. 2.2 Schematic of mouse somite formation. Lateral view of somites budding off of the rostral end of the presomitic mesoderm demonstrates the stepwise transition of mesenchymal cells to epithelium. By convention, the forming somite is labeled "0" and the newest somite is "+1"

In contrast to boundary formation, signals from the surface ectoderm are required to induce MET and the expression of paraxis (Duband et al. 1987, Sosic et al. 1997, Correia and Conlon 2000, Sato et al. 2002, Sato and Takahashi 2005, Linker et al. 2005). Wnt signaling has been implicated in regulating this process with Wnt6 and Wnt11 as the most likely candidates (Wagner et al. 2000, Linker et al. 2005, Geetha-Loganathan et al. 2006, Schmidt et al. 2004). Ectopic expression of *Wnt6* is able to rescue somite epithelialization where the ectoderm has been removed. Further, *Wnt6* is able to induce paraxis transcription in a beta-catenin-dependent manner, predicting a mechanism of action (Linker et al. 2005).

Somite epithelialization is associated with an increase in the expression of members of the cadherin superfamily and cell adhesion molecules (Duband et al. 1987, Tam and Trainor 1994). These cell surface molecules participate in the formation of focal adhesion and desmosomes at the apical junction of epithelium. Inactivation of N-cadherin (*Cdh2*), alone or in combination with cadherin 11 (*Cdh11*), leads to the disorganization of the somite epithelium into small clusters of cells (Radice et al. 1997, Horikawa et al. 1999). The phenotype of the cadherin mutations is not as severe as either the paraxis or *Pax3*, predicting that additional factors associated with cell adhesion are required for epithelialization. The most likely candidates are the genes involved in cytoskeletal remodeling. Likely targets are members of the Rho family of GTPase. In the chick, overexpression of *Cdc42* promotes somitic cells to maintain their mesenchymal state (Nakaya et al. 2004). Both the inhibition and the over activation of Rac1 disrupt somite epithelialization, demonstrating the sensitivity of the cells to disruption of this pathway. The activity of Rac1 cannot be rescued by paraxis, predicting that *Rac1* is acting downstream (Nakaya et al. 2004).

Rostral/Caudal Polarity of Somites

Spatial identity along the rostral/caudal axis is established in each somite at the time of its formation (Aoyama and Asamoto 1988). Rostral/caudal polarity is essential for imposing the segmental patterning of the peripheral nerves and the resegmentation of the sclerotome during vertebrae formation. This is regulated by an intricate feedback loop between cells in the rostral and caudal halves of the forming somite (somite 0). Consistent with the cyclical nature of somitogenesis, the feedback loop is also entrained with the oscillating segmental clock. Activation of the Notch pathway plays a central role in determining spatial identity. Disruption of *Notch1*, ligands *Dll1* and *Dll3*, or modifying gene peptide-*O*-fucosyltransferase 1 (*Pofut1*) and presenilin-1 leads to the loss of rostral- and caudal-specific gene expression, the fusion of the vertebrae, and the segmental pattern of the peripheral nerve pattern (Swiatek et al. 1994, Conlon et al. 1995, Oka et al. 1995, de la Pompa et al. 1997, Hrabe de Angelis et al. 1997, Kusumi et al. 1998, Barrantes et al. 1999, Koizumi et al. 2001, Dunwoodie et al. 2002, Schuster-Gossler et al. 2009). Spatial identity of the rostral half of the somite requires the expression of *Mesp2*, which is transcribed in a broad domain that encompasses presumptive somite –1 before becoming restricted to the rostral half of the presumptive somite (somite 0) (Saga et al. 1997, Takahashi et al. 2000). Mouse embryos deficient in *Mesp2* lead to expanded expression of caudal-specific genes and fused vertebrae. Transcription of *Mesp2* is upregulated by activated Notch in a *Tbx6*-dependent manner (Yasuhiko et al. 2006), which in turn represses transcription of the *Dll1* ligand in the rostral domain through the transcriptional repressor, ripply2 (Morimoto et al. 2007). In the caudal half of somite 0, *Mesp2* transcription is repressed in a presenilin-1-dependent manner (Koizumi et al. 2001, Takahashi et al. 2003, Yasuhiko et al. 2006).

Maintenance of rostral/caudal polarity after somite formation requires paraxis, which is associated with the regulation of somite epithelialization (Johnson et al. 2001). In *paraxis*-null embryos, the transcription pattern of *Mesp2* and components of the Notch signaling pathway are unaltered in somite 0 and –1. However, the expression of caudal-specific genes, such as *Dll1* and Uncx4.1 (*Uncx4.1*), is broadly transcribed in the newly formed somites. It has been proposed that paraxis participates in a cell adhesion-dependent mechanism of maintaining the intersomitic boundary between the rostral and the caudal halves of the somite after their specification in the PSM (Johnson et al. 2001).

The Anatomy and Development of the Vertebrae and IV Discs

A typical vertebra consists of two parts: the body and the vertebral (or neural) arch (Fig. 2.3a). The vertebral body is located anteriorly and articulates with the adjacent IV discs (Figs. 2.1, 2.3, and 2.4). Together, the vertebral body and the arch form a central, vertebral foramen, and, collectively, the foramina create a vertebral canal, protecting the spinal cord. In this section, the functional anatomy of the vertebrae and IV discs in the adult and the genetic basis for their development in the embryo will be discussed.

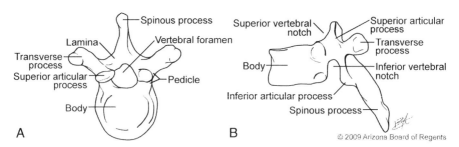

Fig. 2.3 Features of a typical human vertebra. **a**. Superior and **b**. lateral view. Drawing by Brent Adrian

Functional Anatomy of the Vertebrae and IV Discs

The vertebral bodies consist of a shell of compact bone surrounding a core of trabecular bone and red marrow. In addition, hyaline cartilage forms vertebral end plates on the superior and inferior surfaces of each body. The vertebral bodies, in conjunction with the IV discs, bear and transmit weight; as a result, the bodies increase in size from the cervical to the lumbar region (Fig. 2.1). However, as weight is then transferred to the lower extremities via the sacrum, the bodies subsequently decrease in size.

The vertebral arch is located posterior to the vertebral body and consists of two pedicles and two laminae (Fig. 2.3a). The superior and inferior notches of adjacent pedicles form the intervertebral foramina, which transmit the spinal nerves (Figs. 2.1 and 2.3b). Disruption of these foramina (e.g., by a herniated disc) can compress the spinal nerves, leading to both sensory and motor deficits. In addition to protecting the spinal cord and spinal nerves, the vertebral arch also has a number of processes that provide sites for muscle and ligament attachment. The spinous processes, located at the junction of the laminae, and the transverse processes, located at the pedicle–lamina junctions, provide attachment sites for ligaments as well as the erector spinae and transversospinalis muscle groups (Fig. 2.3a, b). In addition, the transverse processes articulate with the costal tubercles to form the costovertebral joints. Finally, the superior and the inferior articular processes of adjacent vertebrae interlock to form the zygapophysial (or facet) joints (Fig. 2.4). These synovial joints permit gliding movements and their orientation largely determines the ranges of motion that are possible between adjacent vertebrae.

The morphology and the functions of the vertebrae vary by region. The cervical spine is composed of seven vertebrae (Fig. 2.1). The bodies are small, reflecting their relatively minor weight-bearing role, while transverse foramina are present for the passage of the vertebral arteries and veins. In addition, the articular facets on the superior and the inferior articular processes face superiorly and inferiorly,

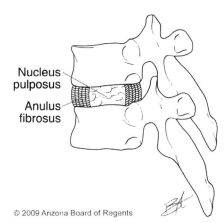

Nucleus
pulposus

Anulus
fibrosus

Fig. 2.4 Structure of the intervertebral disc. Drawing by Brent Adrian   © 2009 Arizona Board of Regents

promoting flexion, extension, lateral flexion, and rotation at the cervical facet joints. This region also includes two highly derived elements, the C1 and C2 vertebrae. The C1 vertebra, or atlas, lacks a body and spinous process. Instead, it features two lateral masses united by an anterior and a posterior vertebral arch. The superior articular facets of the atlas articulate with the occipital condyles of the skull to form the atlanto-occipital joints. These synovial joints allow for flexion and extension of the head. The C2 vertebra, or axis, features a dens or an odontoid process; this process represents the body of the atlas that fuses with the axis during development. The dens process articulates with the anterior arch of the atlas to form the median atlanto-axial joint, while the facet joints between the C1 and C2 vertebrae form the lateral atlanto-axial joints. Together, these joints allow for rotation of the head.

The 12 thoracic vertebrae are distinct in featuring costal facets on their bodies and transverse processes (Fig. 2.3b). Typically, a thoracic vertebral body articulates with two costal heads, while the transverse process articulates with the tubercle of one of these ribs; altogether, these articulations form the costovertebral joints. These synovial joints serve to elevate and depress the ribs, thus increasing the anterior–posterior and transverse diameters of the thoracic cavity during respiration. In the thoracic spine, the superior and inferior articular facets face anteriorly and posteriorly (Fig. 2.3b), permitting rotation and some lateral flexion. However, the orientation of these facets, as well as the inferiorly directed spinous processes and the costovertebral joints, severely restricts flexion and extension of the thoracic spine. In contrast, the medially and laterally facing articular facets of the five lumbar vertebrae allow for a great deal of flexion and extension, but restrict rotation. The lumbar vertebrae also exhibit robust vertebral bodies and well-developed spinous, transverse, and superior articular processes that provide attachment sites for ligaments as well as the erector spinae and transversospinalis muscle groups (Fig. 2.1).

The sacrum is typically formed by the fusion of five sacral vertebrae (Fig. 2.1). The sacral canal transmits the spinal roots of the cauda equina and ends at the sacral hiatus, an important landmark for administering a caudal epidural. In addition, four pairs of sacral foramina transmit the ventral and dorsal rami of the sacral spinal nerves. The sacrum plays an important role in transmitting the weight of the body from the spine to the lower extremities; as a result, the sacroiliac joints are protected by extremely robust ligaments. Similar to the sacrum, the coccyx is typically formed by the fusion of four coccygeal vertebrae (Fig. 2.1). Although the coccyx is rudimentary in humans, it serves as a focal point for the attachment of the muscles of the pelvic floor as well as the sacrotuberous and sacrospinous ligaments.

Most of the vertebral bodies articulate superiorly and inferiorly with IV discs, forming secondary cartilaginous joints or symphyses (Fig. 2.4). However, an IV disc is not present between the atlas and the axis, and the sacral and coccygeal IV discs ossify progressively into adulthood. Representing up to 25% of the total length of the spine, the IV discs act as shock absorbers and enhance spinal flexibility, particularly in the cervical and lumbar regions (Moore and Dalley 2006). The IV discs are responsible for resisting compressive loads due to weight bearing as well as tensile and shearing stresses that arise with movements of the vertebral column, such as rotation and lateral flexion. The thoracic IV discs are relatively thin and uniform in shape, while the cervical and lumbar IV discs are wedge-shaped, contributing to the curvatures of the vertebral column (Fig. 2.1). Each IV disc is composed of an outer fibrocartilaginous ring, the anulus fibrosus, and a central gelatinous core, the nucleus pulposus (Fig. 2.4). Composed primarily of collagen fibers, the anulus fibrosus is characterized by a series of concentric layers, or lamellae (Fig. 2.4). The lamellae serve to resist the expansion of the nucleus pulposus during compression (Cailliet 1988). The nucleus pulposus is composed of water, proteoglycans, and scattered collagen fibers.

The vertebrae and IV discs are stabilized by robust spinal ligaments which function to restrict movements and to minimize the need for continual muscular contraction. The major spinal ligaments are illustrated in Fig. 2.5. The broad anterior longitudinal ligament is situated on the anterior surface of the vertebral bodies and IV discs and extends from the sacrum to the occipital bone (Fig. 2.5). This ligament prevents hyperextension of the spine and anterior herniation of the nucleus pulposus. This ligament is especially prone to injury in the cervical region due to whiplash (hyperextension) injuries. The posterior longitudinal ligament is slender compared to its counterpart. It lies within the vertebral canal, on the posterior surface of the vertebral bodies and IV discs (Fig. 2.5). This ligament prevents hyperflexion of the vertebral column and posterior herniation of the nucleus pulposus. In fact, due to the presence of the posterior longitudinal ligament, the nucleus pulposus tends to herniate in a posterolateral direction.

While the anterior and posterior longitudinal ligaments traverse the length of the spine, the ligamenta flava connect the laminae of adjacent vertebrae (Fig. 2.5). These ligaments contribute to the posterior wall of the vertebral canal, thus helping to protect the spinal cord. The ligamenta flava are highly elastic; they support

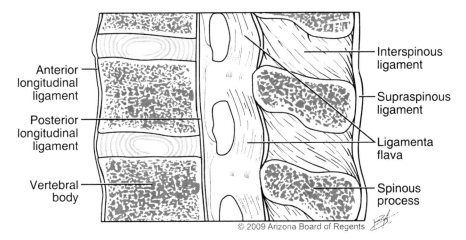

Fig. 2.5 Major ligaments of the spine. Lateral view illustrating the ligamentum flava, supraspinous, interspinous, and anterior and posterior longitudinal ligaments. Drawing by Brent Adrian

the normal curvatures of the spine, resist separation of the laminae during flexion, and assist in extending the spine from a flexed position. The vertebrae are also held together by the intertransverse and interspinous ligaments, which connect adjacent transverse and spinous processes, respectively (Fig. 2.5). More superficially, the robust supraspinous ligament binds the spinous processes together. In the neck, the supraspinous ligament merges with the ligamentum nuchae, a fibroelastic structure that extends from the cervical spinous processes to the occiput, forming a midline raphe for muscle attachment. The intertransverse, interspinous, and supraspinous ligaments help prevent hyperflexion and extreme lateral flexion of the vertebral column.

Development of the Vertebrae

The axial skeleton is derived from the sclerotome compartment of the somites, which first appear during the fourth week of development in humans as the epithelial cells in the ventral/medial quadrant of the somite undergo an epithelial-to-mesenchymal transition (EMT). These cells, in combination with the mesenchymal cells of the somitocoele, form the sclerotome (Fig. 2.6) (reviewed in Dockter 2000). Vertebrae are formed through the process of endochondral ossification. As such, a chondrogenic model of the vertebrae and ribs is developed from sclerotomal cells. Replacement of the cartilage with bone then follows. The molecular events that regulate this process are common with the appendicular and part of the cranial skeleton. These pathways are reviewed elsewhere (Mackie et al. 2008). In this chapter, we will focus on the signaling events that influence patterning of the newly formed vertebrae.

The transition from sclerotome to vertebrae can be divided into (1) the vertebral body and the intervertebral disc at the ventral midline, (2) lateral neural arches that will give rise to the pedicles and transverse process, and (3) the dorsal spinous process. Patterning of the vertebrae along the dorsal/ventral axis is controlled by opposing gradients derived from the notochord and the surface ectoderm overlying the neural tube. Sonic hedgehog (SHH) and the BMP inhibitor, noggin, have been identified as factors expressed in the notochord that are sufficient to promote the expression of the transcription factors *Pax1, Pax9,* and *Mfh1* in the sclerotome (Fan and Tessier-Lavigne 1994, Peters et al. 1995, McMahon et al. 1998, Furumoto et al. 1999). *Pax1* and *Pax9* are essential for the maintenance of sclerotomal cells (Furumoto et al. 1999). Compound mutations of these two genes in the mouse lead to loss of the vertebral body and proximal ribs (Peters et al. 1999). In addition to signals from the notochord, the Polycomb genes *Pbx1* and *Pbx2* and bHLH genes paraxis and *Mesp2* are also required for *Pax1* and *Pax9* transcription (Takahashi et al. 2007, Capellini et al. 2008). The homeodomain-containing genes *Meox1* and *Meox2* have also been implicated in vertebrae development. *Meox1*-deficient mice display segmental fusions in the occipital/cervical region and the *Meox1/Meox2* compound null mutations lead to a profound loss of vertebral structures (Mankoo et al. 2003, Skuntz et al. 2009).

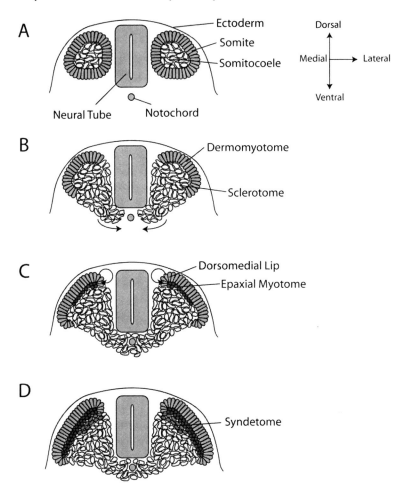

Fig. 2.6 Schematic of somite differentiation. **a.** Newly formed somites consist of an epithelium and a mesenchymal somitocoele. **b.** The sclerotome compartment forms as the medial/ventral region of the somite undergoes MET. **c.** The epaxial myotome forms as epithelial cells of the dorsomedial lip of the dermomyotome undergo MET and migrate subjacently. **d.** The syndetome appears as cells from the myotome interact with dorsal sclerotomal cells

The dorsal elements of the vertebrae, including the spinous process and the dorsal part of the neural arches, require *Msx1* and *Msx2* transcription induced by BMP2 and BMP4 expressed in the surface ectoderm and the roof plate of the neural tube (Monsoro-Burq et al. 1994, 1996, Watanabe et al. 1998). SHH and the BMPs are mutually antagonistic in their actions (Pourquié et al. 1993). Ectopic expression of BMP2 or BMP4 on the dorsal neural tube will increase dorsal chondrogenesis, while ectopic expression lateral to the neural tube inhibits chondrogenesis (Tonegawa et al. 1997, Watanabe et al. 1998). The corollary is also true with SHH-expressing cells grafted dorsally, inhibiting *Msx1* transcription and preventing chondrogenesis (Watanabe et al. 1998).

Remodeling the Sclerotome into Vertebrae

The formation of the vertebrae is dependent on the highly coordinated migration of sclerotomal cells both toward the midline and along the rostral/caudal axis (Fig. 2.6) (reviewed in Brand-Saberi and Christ 2000). Soon after EMT, cells from the ventral/medial sclerotome migrate toward the notochord, where they will contribute to the vertebral body and intervertebral discs. This is followed by the migration of the lateral sclerotomal cells dorsally to form the vertebral pedicles and the laminae of the neural arches. At this point, cells of the rostral and caudal halves of the sclerotome can be distinguished visually based on their density. The caudal half of the sclerotome will proliferate and migrate toward the rostral domain of the adjacent somite (Fig. 2.7a). Fate mapping in chick embryos revealed that the caudal and rostral sclerotome halves of adjacent somites contribute equally to the vertebral body (Goldstein and Kalcheim 1992, Aoyama and Asamoto 2000). In contrast, the neural arches are derived almost solely from the caudal domain and the spinous process from the rostral domain.

Resegmentation of the sclerotome is intimately linked to the specification of the rostral and caudal domains early in somitogenesis. As described previously, the interaction between the Notch signaling pathway and *Mesp2* leads to the specification of the rostral and caudal fate of the somite prior to overt segmentation. As such, the caudalization of the somite by inactivation of *Mesp2* leads to fusion of the vertebral bodies and neural arches along the length of the vertebral column (Saga et al. 1997). In contrast, disruption of the caudal identity of somites through inactivation of the Notch pathway leads to fused vertebral bodies and an absence of neural arches. As will be discussed in detail in later chapters in this book, mutations in genes regulating this process have been identified as the cause of spondylocostal dysostoses, a heterogeneous group of disorders with severe axial skeletal malformation characterized radiographically by multiple vertebral segmentation defects (reviewed in Sparrow et al. 2007). Disruption of rostral/caudal polarity after somite formation has also been shown to impact resegmentation, though to a lesser extent. In *paraxis*-deficient embryos, ventral cartilage fails to segment into vertebral bodies and IV discs, while the lateral neural arches are unaffected (Johnson et al. 2001).

Rostral/Caudal Patterning

An additional layer of regulation is required to confer the distinctive regional characteristics of the cervical, thoracic, lumbar, sacral, and caudal vertebrae. Members of the Hox transcription factor family have been strongly implicated in establishing positional identity of vertebrae along the rostral/caudal axis (reviewed in Wellik 2007). From classic studies in *Drosophila*, the *Hox* genes have long been known to regulate segmental identity in the insect body plan (Lewis 1978). Compound mutations that inactivate more than one gene of a paralogous Hox group in mice lead to rostral homeotic transformation of the vertebrae. This was first observed

A

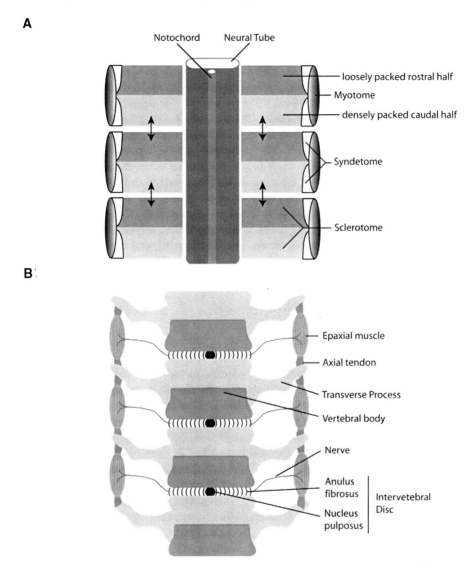

Fig. 2.7 Schematic of vertebral generation through sclerotome resegmentation. **a**. Ventral view of the sclerotome, syndetome, and myotome compartments. The caudal half of the sclerotome grows into the rostral half of the adjacent somite. **b**. Ventral view of the vertebral column with associated epaxial muscles and axial tendons. *Shading* represents the contribution of the rostral and the caudal sclerotomes to the vertebral bodies and transverse processes. The intervertebral disc forms at the site of sclerotome separation. Note the relationship of the muscle and bone after resegmentation

with *Hoxa3/Hoxd3* double mutant embryos, where the prevertebral elements that normally contribute to the atlas form a bone contiguous with the occipital bone (Condie and Capecchie 1994). Since this observation, similar homeotic transformations have been reported for paralogous mutations in the *Hox5*, *Hox6*, *Hox7*,

Hox8, *Hox9*, *Hox10*, and *Hox11* group genes (Chen et al. 1998, van den Akker et al. 2001, Wellik and Capecchi 2003, McIntyre et al. 2007). Consistent with the co-linear expression of these genes, the rostral homeotic transformations effect successively more caudal vertebrae, with the Hox11 paralogous mutants displaying a transformation of sacral and early caudal vertebrae into a lumbar-like fate (Wellik and Capecchi 2003).

The positional identity conferred by the *Hox* genes during vertebrae patterning is modified by members of the polycomb family and TALE class of homeodomain-containing transcription factors. The polycomb genes *Bmi* and *Eed* function as transcriptional repressors that limit the rostral transcription boundary of individual *Hox* genes. Inactivation of these genes leads to a rostral shift in gene expression and transformation of the vertebrae (Kim et al. 2006). The TALE gene families, *Pbx* and *Meis* genes, are able to form dimer partners with the *Hox* genes, leading to modified transcription of target genes by altering DNA-specific binding specificity (reviewed in Moens and Selleri 2006). The TALE genes play a larger role in patterning and regulating the transcription of the 5′ *Hox* genes in both a Hox-dependent and Hox-independent manner (Popperl et al. 1995, Maconochie et al. 1997, Berkes et al. 2004, Capellini et al. 2006).

Formation of the IV Discs

The genesis of the IV discs is intimately linked to somite polarity and sclerotome resegmentation and as such is dependent on the Notch/*Mesp2* signaling (Teppner et al. 2007). The annuli fibrosi of the IV discs forms from condensed mesenchyme derived from the somitocoele at the border of the rostral and caudal domains during resegmentation (Huang et al. 1996, Mittapalli et al. 2005). Somitocoele cells cannot be replaced by sclertomal cells derived from EMT in forming the IV disc predicting specification of a distinct lineage, now called the arthrotome (Mittapalli et al. 2005). Development of the annuli fibrosi and its maintenance in adults is dependent on members of the TGF-beta superfamily. Inactivation of *TGF-beta type II receptor* (*Tgfbr2*) in type II collagen-expressing cells results in an expansion of *Pax1/Pax9* expression and the loss of IV discs (Baffi et al. 2006). GDF-5 and BMP-2 promote cell aggregation and expression of the chondrogenic genes instead of osteogenic genes in the IV discs (Kim et al. 2003, Yoon et al. 2003, Li et al. 2004). The nucleus pulposus is derived from an expansion of notochord cells that move out of the cartilage of the adjacent vertebral body (Paavola et al. 1980). Survival of the notochord cells is dependent on the expression of *Sox5* and *Sox6* (Smits and Lefebvre 2003).

The Anatomy and Development of Spinal Muscles

The spinal muscles function to stabilize and achieve movements of the vertebral column. A number of muscle groups act on the spine. Those located anterior to the vertebral bodies act as flexors. These include longus capitis and colli, psoas major, and rectus abdominis. Lateral flexion is achieved by the scalenes in the cervical region and quadratus lumborum, transversus abdominis, and the abdominal obliques in the lumbar region. The flexors and lateral flexors of the spine are innervated by the ventral rami of spinal nerves. In contrast, the extensors of the spine are located posterior to the vertebral bodies and are innervated by the dorsal rami of spinal nerves (Fig. 2.5). The term "spinal muscles" typically refers to the extensors of the spine. In this section, the functional anatomy of the spinal muscles and the genetic basis for their development in the embryo will be discussed.

Functional Anatomy of the Spinal Muscles

Splenius capitis and cervicis occupy the posterior aspect of the cervical region, deep to trapezius and the rhomboids (Fig. 2.8a). They take origin from the ligamentum nuchae and cervical and thoracic spinous processes and insert onto the mastoid

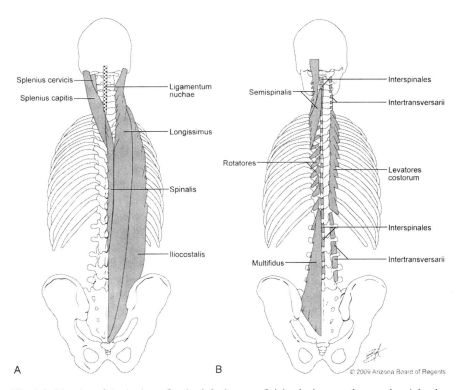

Fig. 2.8 Muscles of the back. **a.** On the *left*, the superficial splenius muscles; on the *right*, the erector spinae muscles, including iliocostalis, longissimus, and spinalis. **b.** On the *left*, the transversospinalis muscles, including semispinalis, multifidus, and rotatores; on the *right*, the levatores costarum, intertransversarii, and interspinales muscles. Drawing by Brent Adrian

process and occipital bone (capitis) or the cervical transverse processes (cervicis) (Fig. 2.8a). Bilateral contraction of splenius capitis and cervicis extends the head and the cervical spine, while unilateral contraction laterally flexes and rotates the neck to the ipsilateral side.

Lying deep to the splenius layer, the erector spinae consist of three longitudinal columns of muscle (Fig. 2.8a). These muscles arise via a common tendon from the iliac crest, sacrum, and lumbar spinous processes. From lateral to medial, the columns include (1) iliocostalis, which attaches to the ribs and cervical transverse processes, (2) longissimus, which attaches to the ribs, thoracic and cervical transverse processes, and mastoid process, and (3) spinalis, which spans adjacent spinous processes. Unilateral contraction of the erector spinae muscles laterally flexes and rotates the spine to the ipsilateral side, while bilateral contraction extends the spine.

The transversospinalis muscles lie deep to the erector spinae. These muscles occupy the region between the transverse and the spinous processes, and include the semispinalis, multifidus, and rotatores muscles (Fig. 2.8b). The semispinalis muscles are located in the thoracic and cervical regions, while the rotatores are prominent in the thoracic region. In contrast, the multifidus extends along the length of the spine but is most developed in the lumbar region. Unilateral contraction of the transversospinalis muscles laterally flexes the spine and rotates it to the contralateral side, while bilateral contraction extends the spine. These muscles stabilize adjacent vertebrae and may have a proprioceptive function (Buxton and Peck 1989, Moore and Dalley 2006).

Deep to the erector spinae are the levatores costarum, intertransversarii, interspinales, and the muscles of the suboccipital triangle (Fig. 2.8b). The levatores costarum are located between the transverse processes and the ribs and act as accessory muscles of respiration. The intertransversarii and the interspinales span the transverse and spinous processes, respectively, and help stabilize the spine. Finally, among the muscles of the suboccipital triangle, the rectus capitis posterior major and minor and the superior oblique extend the atlanto-occipital joints, while the inferior oblique rotates the atlanto-axial joints.

The extensor muscles of the spine may contribute to either the initiation or the progression of scoliotic curves (Fidler and Jowett 1976, Meier et al. 1997, Mannion et al. 1998, Chan et al. 1999). Asymmetry of the spinal extensors, especially the multifidus muscle, has been reported in individuals with idiopathic scoliosis (Zuk 1962, Butterworth and James 1969, Fidler and Jowett 1976, Spencer and Zorab 1976, Alexander and Season 1978, Yarom and Robin 1979, Khosla et al. 1980, Reuber et al. 1983, Sahgal et al. 1983, Zetterberg et al. 1983, Ford et al. 1984, Bylund et al. 1987, Meier et al. 1997, Chan et al. 1999). Chan and colleagues (1999) analyzed MRI data for adolescents with idiopathic scoliosis and found multifidus abnormalities on the concave side of scoliotic curves. As multifidus achieves ipsilateral lateral flexion, this finding suggests that increased contractility or reduced fiber length may play a role in the development of a curvature. The documented increase in type II muscle fibers on the concave side of scoliotic curves, indicating a longer T2 relaxation time, could also contribute to the development of a lateral curve (Ford et al. 1984, Meier et al. 1997).

Development of Spinal Muscles

The spinal muscles that function to stabilize and achieve movements of the vertebral column are derived from the dorsal half of the myotome, from the occipital, thoracic, lumbar, and sacral somites. The origins of spinal muscles lie within a highly mitogenic myogenic progenitor cell (MPC) population located in the dorsomedial margin of the dermomyotome. These cells migrate subjacently to a space between the dermomyotome and the sclerotome where they exit the cell cycle and differentiate into mononucleated myocytes (Fig. 2.6, Ordahl and Le Douarin 1992, Denetclaw et al. 1997). The myotome expands along both the medial/lateral and the dorsal/ventral axes by successive waves of MPC migration from the dermomyotome (Denetclaw et al. 1997, Kahane et al. 1998, Denetclaw and Ordahl 2000, Ordahl et al. 2001). This is followed by fusion of the myocytes into the multinucleated myotubes and morphogenic remodeling into the pattern of the adult spinal muscles (Venters et al. 1999).

The genetic basis of skeletal muscle development has been an area of intense study. The myogenic bHLH transcription factor family including MyoD (*Myod1*), myf-5 (*Myf5*), myogenin (*Myog*), and MRF4 (*Myf6*) have been shown to be essential to initiate and maintain the myogenic program in cells fated to the myogenic lineage. The phenotypes of individual and compound null mutants reveal that these factors can be split into a specification subclass (myf-5 and MyoD) and a differentiation subclass (myogenin and MRF4). Interaction between the myogenic bHLH factors and members of the myocyte enhancer factor-2 (MEF2) family of MADS-box transcription factors enhances muscle differentiation by increasing affinity of DNA binding and expanding the number of target genes that can be activated (reviewed in Molkentin and Olson 1996, Arnold and Braun 2000). The activity of Mef-2 and the myogenic factors are controlled in part by their association with chromatin remodeling proteins histone acetyltransferases (HATs) and histone deacetylases (HDACs) that promote and repress muscle-specific transcription, respectively. Calcium/calmodulin-dependent protein kinase (CaMK)-dependent phosphorylation of HDAC5 leads to its dissociation with MEF-2 and transport out of the nucleus (McKinsey et al. 2000, 2001). Acetylation of MyoD and myf-5 through p300 or PCAF increases affinity of the transcription factors for its DNA target and promotes transcription of myogenin and MRF4 as well as induces cell cycle arrest (Puri et al. 1997, Sartorelli et al. 1999).

Specification of MPCs within the somite fated to become the epaxial muscles is dependent on paracrine factors secreted by adjacent tissues. These signals direct the competence of the cells to initiate the myogenic program and promote the amplification of these committed progenitor cells in the dorsal/medial lip of the dermomyotome. Because of its role in specification, initiating *Myf5* transcription has been used as a readout of specification. A combination of sonic hedgehog (*Shh*) secreted from the notochord and Wnts from the dorsal neural tube and the surface ectoderm is implicated in this process (Reshef et al. 1998, Cossu and Borello 1999, Borycki et al. 2000). Based on explant experiments, Wnt1 is able to induce the transcription of *Myf5* (Tajbakhsh et al. 1998). The activity is transduced by Frizzled receptors 1 and 6 through the canonical β-catenin pathway (Borello et al. 2006). The role of *Shh* in specification was first predicted by the absence of *Myf5* expression in the region of the epaxial myotome in *Shh* null embryos (Borycki et al. 1999). Further, mutations in Gli transcription factors, which transduce Shh signaling, also display a deficit in *Myf5* expression (McDermott et al. 2005). Consistent with these observations, the *Myf5* epaxial enhancer is dependent on a consensus binding sequence for Gli transcription factors and consensus binding sequence for Tcf/Lef, the β-catenin cofactor (Summerbell et al. 2000, Teboul et al. 2003, Borello et al. 2006).

Though the cellular events associated with establishing the early muscle masses, as well as the genetic basis for muscle differentiation, are now well described, less is known about subsequent events associated with establishing individual muscle groups from these masses. Embryonic muscles experience rapid growth, while the early muscles masses in the dorsal body wall, limb, hypoglossal chord, and head undergo several morphological processes (splitting, fusion, directional growth, and movement) in order to establish the appropriate shape, position, and fiber orientation of neonatal muscle. Further, they must coordinate with the growth and differentiation of tendons, ligaments, connective tissue, and skeletal elements to establish the appropriate origin and insertion sites on the bones. Patterning of muscle is dependent on innervation (Yang et al. 2001) and extrinsic signals from the surrounding tissue (Jacob and Christ 1980, Kardon et al. 2003). This is mediated at least in part through mesodermal cells expressing Tcf4 (Kardon et al. 2003) and both intrinsic and extrinsic cues from members of the *Hox* gene family (Ashby et al. 2002, Alvares et al. 2003). However, a clear understanding of the combination of local and global signals that direct individual and functional groups of muscles remains poorly understood.

Tendon Development

The coordinated development of tendons along with muscle and skeletal elements is essential to the proper functioning of the musculoskeletal system. However, the cellular origins of tendons and the regulator pathways that control their specification and differentiation are poorly understood. The recent identification of the bHLH transcription factor, scleraxis, as a tendon-specific marker has accelerated research in this area (Tozer and Duprez 2005). Consistent with its intimate relationship to the epaxial muscles and vertebrae, the axial tendon is derived from a subdomain of the domain of the somite referred to as the syndetome, which is located between the myotome and the sclerotome (Fig. 2.7) (Brent et al. 2003). The syndetomal cells are derived from an interaction between the sclerotome and the myotome. Expression of *Fgf4* and *Fgf8* in the myotome is both necessary and sufficient for scleraxis expression in sclerotomal cells in the future syndetome region (Brent et al. 2003, 2005). Within the sclerotomal cells, the FGF induces an ERK MAP kinase-mediated cascade that requires activation of the ETS transcription factor, *Etv4/Pea3* (Brent and

Tabin 2004, Smith et al. 2005). It appears that there are also inhibitory signals generated from the sclerotome that limit the size of the syndetome. Overexpression of *Pax1* reduces the scleraxis expression domain in the sclerotome; a compound mouse mutation in *Sox5/Sox6* leads to an expansion of the scleraxis-expressing domain (Brent et al. 2005).

Acknowledgments We would like to thank Brent Adrian for preparing Figs. 2.1, 2.3, 2.4, 2.5, and 2.8.

References

Afonin, B., Ho, M., Gustin, J.K., Meloty-Kapella, C., and Domingo, C.R. 2006. Cell behaviors associated with somite segmentation and rotation in Xenopus laevis. Dev Dyn. 235:3268–3279.

Alexander, M.A. and Season, E.H. 1978. Idiopathic scoliosis: an electromyographic study. Arch. Phys. Med. Rehabil. 59:314–315.

Alvares, L.E., Schubert, F.R., Thorpe, C., Mootoosamy, R.C., Cheng, L., Parkyn, G., Lumsden, A., and Dietrich, S. 2003. Intrinsic, Hox-dependent cues determine the fate of skeletal muscle precursors. Dev. Cell 5:379–390.

Aoyama, H. and Asamoto, K. 1988. Determination of somite cells: independence of cell differentiation and morphogenesis. Development. 104:15–28.

Aoyama, H. and Asamoto, K. 2000. The developmental fate of the rostral/caudal half of a somite for vertebra and rib formation: experimental confirmation of the resegmentation theory using chick-quail chimeras. Mech. Dev. 99:71–82.

Arnold, H.H. and Braun, T. 2000. Genetics of muscle determination and development. Curr. Top. Dev. Biol. 48:129–164.

Ashby, P., Chinnah, T., Zakany, J., Duboule, D., and Tickle, C. 2002. Muscle and tendon pattern is altered independently of skeletal pattern in HoxD mutant limbs. J. Anat. 201:422.

Baffi, M.O., Moran, M.A., and Serra, R. 2006. Tgfbr2 regulates the maintenance of boundaries in the axial skeleton. Dev. Biol. 296:363–374.

Barrallo-Gimeno, A., and Nieto, M.A. 2005. The Snail genes as inducers of cell movement and survival: implications in development and cancer. Development 132:3151–3161.

Barrantes, I.B., Elia, A.J., Wünsch, K., Hrabe de Angelis, M.H., Mak, T.W., Rossant, J., Conlon, R.A., Gossler, A., and de la Pompa, J.L. 1999. Interaction between Notch signalling and Lunatic fringe during somite boundary formation in the mouse. Curr Biol. 9:470–480.

Batlle, E., Sancho, E., Franci, C., Dominguez, D., Monfar, M., Baulida, J., and Garcia De Herreros, A. 2000. The transcription factor snail is a repressor of E-cadherin gene expression in epithelial tumour cells. Nat. Cell Biol. 2:84–89.

Berkes, C.A., Bergstrom, D.A., Penn, B.H., Seaver, K.J., Knoepfler, P.S., and Tapscott, S.J. 2004. *Pbx* marks genes for activation by MyoD indicating a role for a homeodomain protein in establishing myogenic potential. Mol. Cell 14:465–477.

Borello, U., Berarducci, B., Murphy, P., Bajard, L., Buffa, V., Piccolo, S., Buckingham, M., and Cossu, G. 2006. The Wnt/beta-catenin pathway regulates Gli-mediated *Myf5* expression during somitogenesis. Development 133:3723–3732.

Borycki, A.M., Brown, A.M., and Emerson, C.P. Jr. 2000. Shh and Wnt signaling pathways converge to control Gli gene activation in avian somites. Development 127:2075–2087.

Borycki, A.G., Brunk, B., Tajbakhsh, S., Buckingham, M., Chiang, C., and Emerson, C.P. Jr. 1999. Sonic hedgehog controls epaxial muscle determination through *Myf5* activation. Development 126:4053–4063.

Brand-Saberi, B., and Christ, B. 2000. Evolution and development of distinct cell lineages derived from somites. Curr. Topics Dev. Biol. 48:1–42.

Brent, A.E., Braun, T., and Tabin, C.J. 2005. Genetic analysis of interactions between the somitic muscle, cartilage and tendon cell lineages during mouse development. Development 132: 515–528.

Brent, A.E., Schweitzer, R., and Tabin, C.J. 2003. A somitic compartment of tendon progenitors. Cell 113:235–248.

Brent, A.E. and Tabin, C.J. 2004. FGF acts directly on the somitic tendon progenitors through the Ets transcription factors Pea3 and Erm to regulate scleraxis expression. Development 131:3885–3896.

Buchberger, A., Seidl, K., Klein, C., Eberhardt, H., and Arnold, H.H. 1998. cMeso-1, a novel bHLH transcription factor, is involved in somite formation in chicken embryos. Dev. Biol. 199:201–215.

Burgess, R., Cserjesi, P., Ligon, K.L., and Olson, E.N. 1995. *Paraxis*: a basic helix-loop-helix protein expressed in paraxial mesoderm and developing somites. Dev. Biol. 168:296–306.

Burgess, R., Rawls, A., Brown, D., Bradley, A., and Olson, E.N. 1996. Requirement of the *paraxis* gene for somite formation and musculoskeletal patterning. Nature 384:570–573.

Butterworth, T.R. and James, C. 1969. Electromyographic studies in idiopathic scoliosis. South Med. J. 62:1008–1010.

Buxton, D.F. and Peck, D. 1989. Neuromuscular spindles relative to joint movement complexities. Clin. Anat. 2:211–224.

Bylund, P., Jansson, E., Dahlberg, E., and Eriksson, E. 1987. Muscle fiber types in thoracic erector spinae muscles. Clin. Orthop. 214:222–228.

Cailliet, R. 1988. Low Back Pain Syndrome. Fourth Edition. Philadelphia: FA Davis Company.

Cano, A., Perez-Moreno, M.A., Rodrigo, I., Locascio, A., Blanco, M.J., del Barrio, M.G., Portillo, F., and Nieto, M.A. 2000. The transcription factor snail controls epithelial-mesenchymal transitions by repressing E-cadherin expression. Nat. Cell Biol. 2:76–83.

Capellini, T.D., Di Giacomo, G., Salsi, V., Brendolan, A., Ferretti, E., Srivastava, D., Zappavigna, V., and Selleri, L. 2006. *Pbx1/Pbx2* requirement for distal limb patterning is mediated by the hierarchical control of Hox gene spatial distribution and Shh expression. Development 133:2263–2273.

Capellini, T.D., Zewdu, R., Di Giacomo, G., Asciutti, S., Kugler, J.E., Di Gregorio, A., and Selleri, L. 2008. *Pbx1/Pbx2* govern axial skeletal development by controlling Polycomb and Hox in mesoderm and *Pax1/Pax9* in sclerotome. Dev. Biol. 321:500–514.

Chan, Y.L., Cheng, J.C.Y., Guo, X., King, A.D., Griffith, J.F., and Metreweli, C. 1999. MRI evaluation of multifidus muscles in adolescent idiopathic scoliosis. Pediatr. Radiol. 29:360–363.

Chen, F., Greer, J., and Capecchi, M.R. 1998. Analysis of Hoxa7/Hoxb7 mutants suggests periodicity in the generation of different sets of vertebrae. Mech. Dev. 77:49–57.

Condie, B.G. and Capecchi, M.R. 1994. Mice with targeted disruptions in the paralogous genes hoxa-3 and hoxd-3 reveal synergistic interactions. Science 370:304–307.

Conlon, R.A., Reaume, A.G., and Rossant, J. 1995. Notch1 is required for the coordinate segmentation of somites. Development. 121:1533–1545.

Correia, K.M. and Conlon, R.A. 2000. Surface ectoderm is necessary for the morphogenesis of somites. Mech. Dev. 91:19–30.

Cossu, G. and Borello, U. 1999. Wnt signaling and the activation of myogenesis in mammals. EMBO J. 18:6867–6872.

Dale, J.K., Malapert, P., Chal, J., Vilhais-Neto, G., Maroto, M., Johnson, T., Jayasinghe, S., Trainor, P., Herrmann, B., and Pourquié, O. 2006. Oscillations of the snail genes in the presomitic mesoderm coordinate segmental patterning and morphogenesis in vertebrate somitogenesis. Dev. Cell 10:355–366.

de la Pompa, J.L., Wakeham, A., Correia, K.M., Samper, E., Brown, S., Aguilera, R.J., Nakano, T., Honjo, T., Mak, T.W., Rossant, J., and Conlon, R.A. 1997. Conservation of the Notch signalling pathway in mammalian neurogenesis. Development 124:1139–1148.

Denetclaw, W.F. Jr., Christ, B., and Ordahl, C.P. 1997. Location and growth of epaxial myotome precursor cells. Development 124:1601–1610.

Denetclaw, W.F. and Ordahl, C.P. 2000. The growth of the dermomyotome and formation of early myotome lineages in thoracolumbar somites of chicken embryos. Development 127: 893–905.

Dockter, J.L. 2000. Sclerotome induction and differentiation. Curr. Top. Dev. Biol. 48:77–127.

Duband, J.L., Dufour, S., Hatta, K., Takeichi, M., Edelman, G.M., and Thiery, J.P. 1987. Adhesion molecules during somitogenesis in the avian embryo. J. Cell Biol. 104:1361–1374.

Dubrulle, J., and Pourquié, O. 2004. Coupling segmentation to axis formation. Development 131:5783–5793.

Dunwoodie, S.L., Clements, M., Sparrow, D.B., Sa, X., Conlon, R.A., and Beddington, R.S. 2002. Axial skeletal defects caused by mutation in the spondylocostal dysplasia/pudgy gene Dll3 are associated with disruption of the segmentation clock within the presomitic mesoderm. Development 129:1795–1806.

Fan, C.M., and Tessier-Lavigne, M. 1994. Patterning of mammalian somites by surface ectoderm and notochord: evidence for sclerotome induction by a hedgehog homolog. Cell 79: 1175–1186.

Fidler, M.W. and Jowett, R.L. 1976. Muscle imbalance in the aetiology of scoliosis. J. Bone Joint Surg. 58-B:200–201.

Ford, D.M., Bagnall, K.M., McFadden, K.D., Greenhill, B.J., and Raso, V.J. 1984. Paraspinal muscle imbalance in adolescent idiopathic scoliosis. Spine 9:373–376.

Furumoto, T.A., Miura, N., Akasaka, T., Mizutanikoseki, Y., Sudo, H., Fukuda, K., Maekawa, M., Yuasa, S., Fu, Y., Moriya, H., Taniguchi, M., Imai, K., Dahl, E., Balling, R., Pavlova, M., Gossler, A., and Koseki, H. 1999. Notochord-dependent expression of MFH1 and PAX1 cooperates maintain the proliferation of sclerotome cells during the vertebral column development. Dev. Biol. 210:15–29.

Geetha-Loganathan, P., Nimmagadda, S., Huang, R., Christ, B., and Scaal, M. 2006. Regulation of ectodermal Wnt6 expression by the neural tube is transduced by dermomyotomal Wnt11: a mechanism of dermomyotomal lip sustainment. Development 133:2897–2904.

Goldstein, R.S. and Kalcheim, C. 1992. Determination of epithelial half-somites in skeletal morphogenesis. Development 116:441–445.

Henry, C.A., Hall, L.A., Burr Hille, M., Solnica-Krezel, L., and Cooper, M.S. 2000. Somites in zebrafish doubly mutant for knypek and trilobite form without internal mesenchymal cells or compaction. Curr. Biol. 10:1063–1066.

Horikawa, K., Radice, G., Takeichi, M., and Chisaka, O. 1999. Adhesive subdivisions intrinsic to the epithelial somites. Dev. Biol. 215:182–189.

Hrab de Angelis, M., McIntyre, J., 2nd, and Gossler, A. 1997. Maintenance of somite borders in mice requires the Delta homologue Dll1. Nature 386:717–721.

Huang, R., Zhi, Q., Neubuser, A., Muller, T.S., Brand-Saberi, B., Christ, B., and Wilting, J. 1996. Function of somite and somitocoele cells in the formation of the vertebral motion segment in avian embryos. Acta Anat. (Basel) 155:231–241.

Jacob, H.J. and Christ, B. 1980. On the formation of muscular pattern in the chick limb. In Teratology of the Limbs. pp. 89–97. Berlin: Walter de Gruyter and Co.

Jiang, Y.J., Aerne, B.L., Smithers, L., Haddon, C., Ish-Horowicz, D., and Lewis, J. 2000. Notch signaling and the synchronization of the somite segmentation clock. Nature 408: 475–479.

Johnson, J., Rhee, J., Parsons, S.M., Brown, D., Olson, E.N., and Rawls, A. 2001. The anterior/posterior polarity of somites is disrupted in paraxis-deficient mice. Dev. Biol. 229: 176–187.

Kahane, N., Cinnamon, Y., and Kalcheim, C. 1998. The cellular mechanism by which the dermomyotome contributes to the second wave of myotome development. Development 125:4259–4271.

Kardon, G., Harfe, B.D., and Tabin, C.T. 2003. A Tcf4-positive mesodermal population provides a prepattern for vertebrate limb muscle patterning. Dev. Cell 5:937–944.

Keynes, R.J. and Stern, C.D. 1988. Mechanisms of vertebrate segmentation. Development 103:413–429.

Khosla, S., Tredwell, S.J., Day, B., Shinn, S.L., and Ovalle, W.K. 1980. An ultrastructural study of multifidus muscle in progressive idiopathic scoliosis-changes resulting from a sarcolemmal defect of the myotendinous junction. J. Neurol. Sci. 46:13–31.

Kim, D.J., Moon, S.H., Kim, H., Kwon, U.H., Park, M.S., Han, K.J., Hahn, S.B., and Lee, H.M. 2003. Bone morphogenetic protein-2 facilitates expression of chondrogenic, not osteogenic, phenotype of human intervertebral disc cells. Spine 28:2679–2684.

Kim, S.Y., Paylor, S.W., Magnuson, T., and Schumacher, A. 2006. Juxtaposed Polycomb complexes co-regulate vertebral identity. Development 133:4957–4968.

Koizumi, K., Nakajima, M., Yuasa, S., Saga, Y., Sakai, T., Kuriyama, T., Shirasawa, T., and Koseki, H. 2001. The role of presenilin 1 during somite segmentation. Development 128: 1391–1402.

Kulesa, P.M. and Fraser, S.E. 2002. Cell dynamics during somite boundary formation revealed by time-lapse analysis. Science 298:991–995.

Kulesa, P.M., Schnell, S., Rudloff, S., Baker, R.E., and Maini, P.K. 2007. From segment to somite: segmentation epithelialization analyzed within quantitative frameworks. Dev. Dyn. 236: 1392–1402.

Kusumi, K., Sun, E.S., Kerrebrock, A.W., Bronson, R.T., Chi, D.C., Bulotsky, M.S., Spencer, J.B., Birren, B.W., Frankel, W.N., and Lander, E.S. 1998. The mouse pudgy mutation disrupts Delta homologue *Dll3* and initiation of early somite boundaries. Nat. Genet. 19(3): 274–278.

Lewis, E.B. 1978. A gene complex controlling segmentation in *Drosophila*. Nature 276:565–570.

Li, J., Yoon, S.T., and Hutton, W.C. 2004. Effect of bone morphogenetic protein-2 (BMP-2) on matrix production, other BMPs, and BMP receptors in rat intervertebral disc cells. J. Spinal Disord. Tech. 17:423–428.

Linker, C., Lesbros, C., Gros, J., Burrus, L.W., Rawls, A., and Marcelle, C. 2005. Beta-Catenin-dependent Wnt signalling controls the epithelial organisation of somites through the activation of *paraxis*. Development 132:3895–3905.

Mackie, E.J., Ahmed, Y.A., Tatarczuch, L., Chen, K.S., and Mirams, M. 2008. Endochondral ossification: how cartilage is converted into bone in the developing skeleton. Int. J. Biochem. Cell Biol. 40:46–62.

Maconochie, M.K., Nonchev, S., Studer, M., Chan, S.K., Popperl, H., Sham, M.H., Mann, R.S., and Krumlauf, R. 1997. Cross-regulation in the mouse HoxB complex: the expression of Hoxb2 in rhombomere 4 is regulated by Hoxb1. Genes Dev. 11:1885–1895.

Mankoo, B.S., Skuntz, S., Harrigan, I., Grigorieva, E., Candia, A., Wright, C.V., Arnheiter, H., and Pachnis, V. 2003. The concerted action of Meox homeobox genes is required upstream of genetic pathways essential for the formation, patterning and differentiation of somites. Development 130:4655–4664.

Mannion, A.F., Meier, M., Grob, D., and Müntener, M. 1998. Paraspinal muscle fibre type alterations associated with scoliosis: an old problem revisited with new evidence. Eur. Spine J. 7:289–293.

McDermott, A., Gustafsson, M., Elsam, T., Hui, C.C., Emerson, C.P. Jr., and Borycki, A.G. 2005. Gli2 and Gli3 have redundant and context-dependent function in skeletal muscle formation. Development 132:345–357.

McIntyre, D.M., Rakshit, S., Yallowitz, A.R., Loken, L., Jeannotte, L., Capecchi, M.R., and Wellik, D.M. 2007. *Hox* Patterning of the vertebrate rib cage. Development 134: 2981–2989.

McKinsey, T.A., Zhang, C.L., Lu, J., and Olson, E.N. 2000. Signal-dependent nuclear export of a histone deacetylase regulates muscle differentiation. Nature 408:106–111.

McKinsey, T.A., Zhang, C.L., and Olson, E.N. 2001. Control of muscle development by dueling HATs and HDACs. Curr. Opin. Genet. Dev. 11:497–504.

McMahon, J.A., Takada, S., Zimmerman, L.B., and McMhaon, A.P. 1998. Noggin-mediated antagonism of BMP signaling is required for growth and patterning of the neural tube and somite. Genes Dev. 12:1438–1452.

Meier, M.P., Klein, M.P., Krebs, D., Grob, D., and Müntener, M. 1997. Fiber transformations in multifidus muscle of young patients with idiopathic scoliosis. Spine 22:2357–2364.

Mittapalli, V.R., Huang, R., Patel, K., Christ, B., and Scaal, M. 2005. Arthrotome: a specific joint forming compartment in the avian somite. Dev. Dyn. 234:48–53.

Moens, C.B. and Selleri, L. 2006. Hox cofactors in vertebrate development. Dev. Biol. 291: 193–206.

Molkentin, J.D. and Olson, E.N. 1996. Defining the regulatory networks for muscle development. Curr. Opin. Genet. Dev. 6:445–453.

Monsoro-Burq, A.H., Bontoux, M., Teillet, M.A., and Le Douarin, N.M. 1994. Heterogeneity in the development of the vertebra. Proc. Natl. Acad. Sci. U.S.A. 91:10435–10439.

Monsoro-Burq, A.H., Duprez, D., Watanabe, Y., Bontoux, M., Vincent, C., Brickell, P., and Le Douarin, N. 1996. The role of bone morphogenetic proteins in vertebral development. Development 122:3607–3616.

Moore, K.L. and Dalley, A.F. 2006. Clinically Oriented Anatomy. Baltimore: Lippincott Williams and Wilkins.

Morimoto, M., Sasaki, N., Oginuma, M., Kiso, M., Igarashi, K., Aizaki, K., Kanno, J., and Saga, Y. 2007. The negative regulation of *Mesp2* by mouse Ripply2 is required to establish the rostro-caudal patterning within a somite. Development 134:1561–1569.

Nakaya, Y., Kuroda, S., Katagiri, Y.T., Kaibuchi, K., and Takahashi, Y. 2004. Mesenchymal-epithelial transition during somitic segmentation is regulated by differential roles of Cdc42 and Rac1. Dev. Cell 7:425–438.

Oka, C., Nakano, T., Wakeham, A., de la Pompa, J.L., Mori, C., Sakai, T., Okazaki, S., Kawaichi, M., Shiota, K., Mak, T.W., and Honjo, T. 1995. Disruption of the mouse *RBP-J kappa* gene results in early embryonic death. Development 121:3291–3301.

Ordahl, C.P., Berdougo, E., Venters, S.J., and Denetclaw, W.F. Jr. 2001. The dermomyotome dorso-medial lip drives growth and morphogenesis of both the primary myotome and dermomyotome epithelium. Development 128:1731–1744.

Ordahl, C.P. and Le Douarin, N.M. 1992. Two myogenic lineages within the developing somite. Development 114:339–353.

Paavola, L.G., Wilson, D.B., and Center, E.M. 1980. Histochemistry of the developing notochord, perichordal sheath and vertebrae in Danforth's short-tail (sd) and normal C57BL/6 mice. J. Embryol. Exp. Morphol. 55:227–245.

Palmeirim, I., Dubrulle, J., Henrique, D., Ish-Horowicz, D., and Pourquié, O. 1998. Uncoupling segmentation and somitogenesis in the chick presomitic mesoderm. Dev. Genet. 23: 77–85.

Peters, H., Doll, U., and Niessing, J. 1995. Differential expression of the chicken Pax-1 and Pax-9 gene: in situ hybridization and immunohistochemical analysis. Dev. Dyn. 203:1–16.

Peters, H., Wilm, B., Sakai, N., Imai, K., Maas, R., and Balling, R. 1999. *Pax1* and *Pax9* synergistically regulate vertebral column development. Development 126:5399–5408.

Popperl, H., Bienz, M., Studer, M., Chan, S.K., Aparicio, S., Brenner, S., Mann, R.S., and Krumlauf, R. 1995. Segmental expression of Hoxb-1 is controlled by a highly conserved autoregulatory loop dependent upon exd/*pbx*. Cell 81:1031–1042.

Pourquie, O., Coltey, M., Teillet, M.A., Ordahl, C., and Le Douarin, M. 1993. Control of dorsoventral patterning of somitic derivatives by notochord and floor plate. Proc. Natl. Acad. Sci. U.S.A. 90:5242–5246.

Puri, P.L., Sartorelli, V., Yang, X.J., Hamamori, Y., Ogryzko, V.V., Howard, B.H., Kedes, L., Wang, J.Y., Graessmann, A., Nakatani, Y., and Levrero, M. 1997. Differential roles of p300 and PCAF acetyltransferases in muscle differentiation. Mol. Cell 1:35–45.

Radice, G.L., Rayburn, H., Matsunami, H., Knudsen, K.A., Takeichi, M., and Hynes, R.O. 1997. Developmental defects in mouse embryos lacking N-cadherin. Dev. Biol. 181:64–78.

Reshef, R., Maroto, M., and Lassar, A.B. 1998. Regulation of dorsal somitic cell fates: BMPs and Noggin control the timing and pattern of myogenic regulator expression. Genes Dev. 12: 290–303.

Reuber, M., Schultz, A., McNeill, T., and Spencer, D. 1983. Trunk muscle myoelectric activities in idiopathic scoliosis. Spine 8:447–456.

Saga, Y., Hata, N., Koseki, H., and Taketo, M.M. 1997. *Mesp2*: a novel mouse gene expressed in the presegmented mesoderm and essential for segmentation initiation. Genes Dev. 11: 1827–1839.

Sahgal, V., Shah, A., Flanagan, N., Schaffer, M., Kane, W., Subramani, V., and Singh, H. 1983. Morphologic and morphometric studies of muscle in idiopathic scoliosis. Acta Orthop. 54: 242–251.

Sartorelli, V., Puri, P.L., Hamamori, Y., Ogryzko, V., Chung, G., Nakatani, Y., Wang, J.Y., and Kedes, L. 1999. Acetylation of MyoD directed by PCAF is necessary for the execution of the muscle program. Mol. Cell 4:725–734.

Sato, Y. and Takahashi, Y. 2005. A novel signal induces a segmentation fissure by acting in a ventral-to-dorsal direction in the presomitic mesoderm. Dev. Biol. 282:183–191.

Sato, Y., Yasuda, K., and Takahashi, Y. 2002. Morphological boundary forms by a novel inductive event mediated by Lunatic fringe and Notch during somitic segmentation. Development 129:3633–3644.

Schmidt, C., Stoeckelhuber, M., McKinnell, I., Putz, R., Christ, B., and Patel, K. 2004. Wnt 6 regulates the epithelialisation process of the segmental plate mesoderm leading to somite formation. Dev. Biol. 271:198–209.

Schubert, F.R., Tremblay, P., Mansouri, A., Faisst, A.M., Kammandel, B., Lumsden, A., Gruss, P., and Dietrich, S. 2001. Early mesodermal phenotypes in splotch suggest a role for *Pax3* in the formation of epithelial somites. Dev. Dyn. 222:506–521.

Schuster-Gossler, K., Harris, B., Johnson, R., Serth, J., and Gossler, A. 2009. Notch signalling in the paraxial mesoderm is most sensitive to reduced *Pofut1* levels during early mouse development. BMC Dev. Biol. 9:6.

Skuntz, S., Mankoo, B., Nguyen, M.T., Hustert, E., Nakayama, A., Tournier-Lasserve, E., Wright, C.V., Pachnis, V., Bharti, K., and Arnheiter, H. 2009. Lack of the mesodermal homeodomain protein *MEOX1* disrupts sclerotome polarity and leads to a remodeling of the cranio-cervical joints of the axial skeleton. Dev. Biol. 2009 Aug 15;332(2):383–95.

Smith, T.G., Sweetman, D., Patterson, M., Keyse, S.M., and Münsterberg, A. 2005. Feedback interactions between MKP3 and ERK MAP kinase control scleraxis expression and the specification of rib progenitors in the developing chick somite. Development 132: 1305–1314.

Smits, P. and Lefebvre, V. 2003. Sox5 and Sox6 are required for notochord extracellular matrix sheath formation, notochord cell survival and development of the nucleus pulposus of intervertebral discs. Development 130:1135–1148.

Sosić, D., Brand-Saberi, B., Schmidt, C., Christ, B., and Olson, E.N. 1997. Regulation of *paraxis* expression and somite formation by ectoderm- and neural tube-derived signals. Dev. Biol. 185:229–243.

Sparrow, D.B., Chapman, G., Turnpenny, P.D., and Dunwoodie, S.L. 2007. Disruption of the somitic molecular clock causes abnormal vertebral segmentation. Birth Defects Res. C Embryo Today 81:93–110.

Spencer, G.S. and Zorab, P.A. 1976. Spinal muscle in scoliosis. Part 1: histology and histochemistry. J. Neurol. Sci. 30:127–142.

Summerbell, D., Ashby, P.R., Coutelle, O., Cox, D., Yee, S., and Rigby, P.W. 2000. The expression of *Myf5* in the developing mouse embryo is controlled by discrete and dispersed enhancers specific for particular populations of skeletal muscle precursors. Development 127: 3745–3757.

Swiatek, P.J., Lindsell, C.E., del Amo, F.F., Weinmaster, G., and Gridley, T. 1994. *Notch1* is essential for postimplantation development in mice. Genes Dev. 8:707–719.

Tajbakhsh, S., Borello, U., Vivarelli, E., Kelly, R., Papkoff, J., Duprez, D., Buckingham, M., and Cossu, G. 1998. Differential activation of *Myf5* and MyoD by different Wnts in explants of mouse paraxial mesoderm and the later activation of myogenesis in the absence of *Myf5*. Development 125:4155–4162.

Takahashi, Y., Inoue, T., Gossler, A., and Saga, Y. 2003. Feedback loops comprising *Dll1, Dll3* and *Mesp2*, and differential involvement of Psen1 are essential for rostrocaudal patterning of somites. Development 130:4259–4268.

Takahashi, Y., Koizumi, K., Takagi, A., Kitajima, S., Inoue, T., Koseki, H., and Saga, Y. 2000. *Mesp2* initiates somite segmentation through the Notch signalling pathway. Nat. Genet. 25:390–396.

Takahashi, Y. and Sato, Y. 2008. Somitogenesis as a model to study the formation of morphological boundaries and cell epithelialization. Develop. Growth Differ. 50:S149–S155.

Takahashi, Y., Takagi, A., Hiraoka, S., Koseki, H., Kanno, J., Rawls, A., and Saga, Y. 2007. Transcription factors *Mesp2* and *Paraxis* have critical roles in axial musculoskeletal formation. Dev. Dyn. 236:1484–1494.

Tam, P.P. and Trainor, P.A. 1994. Specification and segmentation of the paraxial mesoderm. Anat. Embryol. 189:275–305.

Tanaka, M. and Tickle, C. 2004. *Tbx18* and boundary formation in chick somite and wing development. Dev. Biol. 268:470–480.

Teboul, L., Summerbell, D., and Rigby, P.W. 2003. The initial somitic phase of *Myf5* expression requires neither Shh signaling nor Gli regulation. Genes Dev. 17:2870–2874.

Teppner, I., Becker, S., de Angelis, M.H., Gossler, A., and Beckers, J. 2007. Compartmentalised expression of Delta-like 1 in epithelial somites is required for the formation of intervertebral joints. BMC Dev. Biol. 7:68.

Tonegawa, A., Funayama, N., Ueno, N., and Takahashi, Y. 1997. Mesodermal subdivision along the mediolateral axis in chicken controlled by different concentrations of BMP-4. Development 124:1975–1984.

Tozer, S. and Duprez, D. 2005. Tendon and ligament: development, repair and disease. Birth Defects Res. C Embryo Today 75:226–236.

van den Akker, E., Fromental-Ramain, C., deGraaf, W., LeMouellic, H., Brulet, P., Chambon, P., and Deschamps, J. 2001. Axial skeletal patterning in mice lacking all paralogous group 8 Hox genes. Development 128:1911–1921.

Venters, S.J., Thorsteinsdottir, S., and Duxson, M.J. 1999. Early development of the myotome in the mouse. Dev. Dyn. 216:219–232.

Wagner, J., Schmidt, C., Nikowits, W. Jr., and Christ, B. 2000. Compartmentalization of the somite and myogenesis in chick embryos are influenced by wnt expression. Dev. Biol. 228:86–94.

Watanabe, Y., Duprez, D., Monsoro-Burq, A.H., Vincent, C., and Le Douarin, N.M. 1998. Two domains in vertebral development: antagonistic regulation by SHH and BMP4 proteins. Development 125:2631–2639.

Wellik, D.M. 2007. *Hox* patterning of the vertebrate axial skeleton. Dev. Dyn. 236:2454–2463.

Wellik, D.M. and Capecchi, M.R. 2003. Hox10 and Hox11 genes are required to globally pattern the mammalian skeleton. Science 301:363–366.

Wood, A. and Thorogood, P. 1994. Patterns of cell behavior underlying somitogenesis and notochord formation in intact vertebrate embryos. Dev. Dyn. 201:151–167.

Yang, X., Arber, S., William, C., Li, L., Tanabe, Y., Jessell, T.M., Birchmeier, C., and Burden, S.J. 2001. Patterning of muscle acetylcholine receptor gene expression in the absence of motor innervation. Neuron 30:399–410.

Yarom, R. and Robin, G.C. 1979. Studies on spinal and peripheral muscles from patients with scoliosis. Spine 4:12–21.

Yasuhiko, Y., Haraguchi, S., Kitajima, S., Takahashi, Y., Kanno, J., and Saga, Y. 2006. *Tbx6*-mediated Notch signaling controls somite-specific *Mesp2* expression. Proc. Natl. Acad. Sci. U.S.A. 103:3651–3656.

Yoon, S.T., Su Kim, K., Li, J., Soo Park, J., Akamaru, T., Elmer, W.A., and Hutton, W.C. 2003. The effect of bone morphogenetic protein-2 on rat intervertebral disc cells in vitro. Spine 28: 1773–1780.

Zetterberg, C., Aniansson, A., and Grimby, G. 1983. Morphology of the paravertebral muscles in adolescent idiopathic scoliosis. Spine 8:457–462.

Zuk, T. 1962. The role of spinal and abdominal muscles in the pathogenesis of scoliosis. J. Bone Joint Surg. Br. 44:102–105.

Chapter 3
Environmental Factors and Axial Skeletal Dysmorphogenesis

Peter G. Alexander and Rocky S. Tuan

Introduction

Axial skeletal development is part of the complex, inclusive process of axial or midline development. It involves the interaction of many tissues including the embryonic notochord, neural tube, somite compartments, intersomitic angiopotent cells, and neural crest cells. These tissues give rise to the axial skeleton, intervertebral discs, spinal cord, trunk musculature and dorsal dermis, intervertebral arteries, and spinal ganglia. Development of these tissues occurs in an interdependent and hierarchical manner over an extended period of time. These characteristics may make the axial skeleton disproportionately susceptible to environmental influence, accounting for the high incidence of axial skeletal defects among live and stillbirths. It may also account for the many manifestations of axial skeletal defects observed.

Data show that the axial skeleton is one of several organ systems with a high frequency of abnormality, 1 in 1,000 live births (Brent and Fawcett 2007, Cohen 1997, Dias 2007, Erol et al. 2002, Jaskwhich et al. 2000, O'Rahilly and Müller 1996, Oskouian et al. 2007) and a very low heritable component, estimated to be between 0.5 and 2%. Congenital axial skeletal defects may occur in isolation or as a component of more widespread syndromes or sequences (Cohen 1997, Dias, 2007, Erol et al. 2002, Jaskwhich et al. 2000, Oskouian et al. 2007) (Table 3.1 and Chapter 7). It is estimated that the skeletal defect is accompanied by an intra-spinal neural defect in 40% of cases. In addition, approximately 50–60% cases of congenital scoliosis suffer additional congenital defects in other organ systems including urogenital and cardiovascular systems (approximately 20% and 10–12%, respectively), gastrointestinal and limb defects (2–5%). These combinations of congenital defects and their frequencies are reflective of the degree of concurrent development of the different organ systems.

R.S. Tuan (✉)
Center for Cellular and Molecular Engineering, Department of Orthopaedic Surgery, University of Pittsburgh School of Medicine, Pittsburgh, PA 15219, USA
e-mail: rst13@pitt.edu

K. Kusumi, S.L. Dunwoodie (eds.), *The Genetics and Development of Scoliosis*,
DOI 10.1007/978-1-4419-1406-4_3, © Springer Science+Business Media, LLC 2010

Table 3.1 Genetic syndromes that are characterized by scoliosis

Syndrome	Features
Alagille syndrome (autosomal dominant)	Neonatal jaundice, cholestasis, peripheral pulmonic stenosis, occasional septal defects and patent ductus arteriosus, accompanied by abnormal facies, ocular, *vertebral*, and nervous system abnormalities
Bertolotti syndrome	*Sacralization of the fifth lumbar vertebrae* with sciatica and *scoliosis*
Caudal dysgenesis (agenesis, regression) syndrome	Failure *to form part or all of the coccygeal, sacral, and lumbar vertebrae* and corresponding spinal segments with malformation and dysfunction of the bowel and bladder
Cerebrocostomandibular syndrome (autosomal recessive)	Severe micrognathia, *severe costovertebral anomalies* including bell-shaped thorax, incompletely ossified, aberrant rib structure, abnormal rib connection to the vertebral body, accompanied by palatal defects, glossoptosis, pre- and post-natal growth deficiencies, mental retardation
Coffin-Siris syndrome	Hypoplasia of the fifth fingers and toes associated with mental and growth retardation, coarse facies, mild microcephaly, hypotonia, lax joints, mild hirsutism, and occasionally accompanied by cardiac, *vertebral,* and gastrointestinal abnormalities
Oculocerebral hypopigmentation syndrome (autosomal recessive)	Oculocutaneous albinism, microphthalmus, opaque corneas, oligophrenia with spasticity, high-arched palate, gingival atrophy, *scoliosis*
Kabuki make-up syndrome	Mental retardation, dwarfism, *scoliosis*, cardiovascular abnormalities, and facies reminiscent of a Japanese Kabuki actor
King's syndrome (malignant hyperthermia)	Short stature, *kyphoscoliosis*, pectus carinatum, cryptorchidism, delayed motor development, progressive myopathy, structural cardiovascular defects
Klippel–Feil syndrome	Reduced number of cervical vertebrae, *cervical hemivertebrae*, low hair-line, reduced neck mobility
Lenz's syndrome (X-linked)	Microphthalmia, anophthalmia, digital anomalies, narrow shoulders, double thumbs, *vertebral abnormalities*, dental, urogenital, and cardiovascular defects may occur
Multiple pterygium syndrome (autosomal recessive)	Pterygia of the neck, axillae, popliteal, antecubital, and intercrural areas, accompanied by hypertelorism, cleft palate, micrognathia, ptosis, short stature, and a wealth of skeletal anomalies including camptodactyly, syndactyly, equinovarus, rocker-bottom feet, *vertebral fusions, and rib abnormalities*

Table 3.1 (continued)

Syndrome	Features
Oculoauricularvertebral syndrome (Goldenhar syndrome)	Colobomas of the upper eye lids, bilateral accessory auricular appendages, *vertebral anomalies*, facial bossing, asymmetrical skull, low hair-line, mandibular hypoplasia, low-set ears, and sometimes hemifacial microsomia
Rubenstein–Taybi syndrome	Mental and motor retardation, broad thumbs and big toes, short stature, high-arched palate, straight, beaked nose, various eye abnormalities, pulmonary stenosis, keloid formation at surgical scars, large foramen magnum, *vertebral and sternal abnormalities*
Spondylothoracic dysplasia (Jarcho–Levin syndrome) (autosomal recessive)	*Multiple vertebral defects*, short thorax, rib abnormalities, camptodactyly, syndactyly, and accompanied by urogenital anomalies and respiratory dysfunction
VATER-VACTERL sequence	*Vertebral anomalies,* anal atresia, (cardiac abnormalities), tracheal fistula with esophageal atresia, renal defects, (limb abnormalities)

A selected list of recognized genetic syndromes that may include vertebral anomalies. Genetic syndromes associated with scoliosis are further discussed in Chapter 7.

While dramatic axial skeletal defects do occur in the context of syndromes and other anomalies, the majority of congenital spinal anomalies involve single structural defects of the spine and frequently few obvious coincident malformations or functional deficits (Erol et al. 2002, Jaskwhich et al. 2000, Oskouian et al. 2007), indicating that a time-dependent, tissue-specific insult may be involved. The complexity of axial skeletal development and the variety of axial skeletal defects suggest a variety of loci and mechanisms through which environmental factors may cause axial skeletal dysmorphogenesis.

Faced with the high social costs of resultant morbidity, it is critical to determine the possible impact any environmental factor may have on the embryo. Although many of the known human teratogens can produce axial skeletal defects, the etiology of over half of observed axial skeletal defects is unknown and is assumed to be multi-factorial, a combination of genetic susceptibility and environmental insult (Cohen 1997, Jaskwhich et al. 2000). This fact highlights the need for investigating the role of environmental factors, alone or in combination, in the production of this particular class of defects. Such study requires the convergence of at least two broad fields of study. The first is developmental biology, to understand the details of normal development and identify new markers, loci, and perhaps possible mechanisms of teratogenesis. The second field is teratology, a discipline closely related to reproductive toxicology that involves assessing the impact of environmental factors on the new biological markers, loci, and mechanisms discovered and characterized in developmental biology.

Vertebral Dysmorphogenesis in Human Congenital Scoliosis

Clinically, congenital scoliosis is defined as a spinal curvature of over 10% caused by a structural vertebral defect (Dias 2007, Erol et al. 2002, Oskouian et al. 2007). The abnormal spinal curvature is further defined by its anterior–posterior location and the plane of curvature as coronal for scoliosis and sagittal for kyphosis. The characteristic feature of congenital axial skeletal defects is the malformation of vertebral bodies or processes evident at birth. Broadly, these vertebral defects are clinically classified as failures in formation and morphogenesis represented by hemivertebrae, wedge vertebrae, open vertebral arches, bifid vertebrae, and vertebral agenesis or failures in segmentation represented by unilateral unsegmented bars or block vertebrae bilateral fusions (Fig. 3.1) (Dias 2007, Erol et al. 2002, Jaskwhich et al. 2000, Oskouian et al. 2007). Developmentally, however, all of these defects have their origin in somitogenesis, the initial manifestation of the vertebral column's metameric segmentation.

Fig. 3.1 Different forms of congenital scoliosis: block vertebrae (**a**), unilateral bar, (**b**), wedge vertebrae, (**c**), multiple hemivertebrae, (**d**), single, semi-segmented vertebrae, (**e**), non-segmented hemivertebrae, (**f**), incarcerated hemivertebrae, (**g**), defects in segmentation can produce these defects

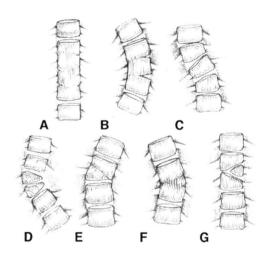

Normal Development of the Axial Skeleton

As discussed in previous chapters, the axial skeleton is derived from the paraxial mesoderm, a primary germ layer, which undergoes the molecularly timed process of somitogenesis to produce blocks of tissue symmetrically arranged on either side of the midline neural tube and notochord (Table 3.2, Fig. 3.1) (Christ et al. 2000, Stockdale et al. 2000, Gridley 2006). The somite is a transient embryonic structure that plays an important role in the patterning of the axial skeleton (comprised of vertebral bodies, ribs, and intervertebral discs) and its associated tissues: the hypaxial and epaxial muscles of the spine, the dorsal dermis of the trunk, and the intervertebral arteries. The morphogenic description of somitogenesis can be conceptually

Table 3.2 Developmental timing of the axial skeleton in the human embryo

Developmental feature	Day of gestation	Other notable occurrences
Gastrulation	15	Neural plate formation
Notochord formation	17–19	Neural tube folding
First somite	19	Heart tube formation
Onset of neural tube fusion	22	Heart tube folding, optic and otic vesicle formation begins
Anterior neuropore closure	23–26	Embryonic circulation
Posterior neuropore closure	26–30	Forelimb bud
Sclerotomal segmentation	24–35	
Notochordal segmentation	28–30	
Last (30th) somite formed	32	Hind limb bud, optic cup formed
All rib primordia evident	42–44	
Chondrification of centra	36–42	
Chondrification of ribs and laminae	40–44	
Chondrification complete/onset of ossification	56–60	

divided into several phases: patterning, morphogenesis, differentiation, and growth and maturation (Tam and Trainor 1994, Christ et al. 2000, Alexander et al. 2007a). These are helpful classifications when characterizing and studying birth defects and their causes.

Among the tissues of the spine, the axial skeleton and its composite tissues undergo multiple rounds of patterning, differentiation, and growth events, including somitogenesis, resegmentation, and ossification, among other processes. Briefly, the axial skeleton is derived from the sclerotome, the ventromedial quadrant of each somite. Cells of sclerotome are initially part of the epithelial somite. Shortly after expressing the paired-box gene *Pax1* (Wallin et al. 1994, Barnes et al. 1996a), the cells de-epithelialize and relocate themselves to surround the notochord. These cells then begin expressing *Sox9*, a chondrocyte-specific transcription factor, and producing prodigious amounts of cartilage matrix to form the cartilage anlage of the vertebral body (Healy et al. 1999). There is a distinct polarity to the somite as it matures (Tam and Trainor 1994) that is consequential in the course of resegmentation, in which the posterior half of one somite merges with the anterior half of the posterior somite (Christ et al. 2004). Together, these halves combine to form a vertebral body out of phase with the other tissue, characteristic of the vertebral motor unit.

Development of the axial skeleton and the surrounding tissues occurs in an interdependent and hierarchical manner over an extended period of time. This may make the axial skeleton disproportionately susceptible to environmental influence, accounting for the high incidence of axial skeletal defects among live and stillbirths. It may also account for the many manifestations of axial skeletal defects observed. Understanding these processes (the normal development of the spine) and their effects upon the surrounding tissues is important in deciphering the etiology of various forms of congenital scoliosis and the mechanisms by which environmental agents may initiate abnormal development.

Experimental Axial Skeletal Teratology

Given that the majority of axial defects have no known genetic cause (Cohen 1997, Dias 2007, Erol et al. 2002, Jaskwhich et al. 2000, Oskouian et al. 2007), the assumption must be made that there is an environmental component. The principles that helped define environmental teratogical agents were popularized in the wake of the "thalidomide experience" and with some modification remain applicable today (Wilson 1977, Sulik 1997, Sadler and Hunter 1994). In establishing the role of an environmental agent in inducing a congenital axial skeletal defect, we know that it must first affect the development and function of axial tissues and those that influence their differentiation including the notochord, neural tube, paraxial mesoderm, and overlying ectoderm. Second, the exposure must occur somewhere between the 4th and 10th week of human gestation, or organogenesis, during which time gastrulation, neurulation, and somitogenesis occur (Table 3.2) (Nishimura et al. 1974, O'Rahilley and Müller 1996). Third, the target of the teratogen must play a necessary role in the affected developmental process (e.g., somitogenesis) by acting via a specific mechanism. Finally, we must observe the dose-response effect of the environmental agent on embryonic development in both frequency and degree of malformation, including the graded manifestations of abnormal development: death, dysmorphogenesis, inhibition of growth or developmental delay, and functional deficit.

In the etiology of scoliosis, target organs may include the paraxial mesoderm and somites, the neural tube and notochord, and the overlying ectoderm. In particular, the patterning of the somite boundaries and the subsequent boundaries of differentiation defined by integrated signaling pathways under the influence of the surrounding tissues figures prominently. Morphological processes that may be affected include somitogenesis, neurulation, and gastrulation, which involve cell migration, epithelialization, and laminar fusion, as well as proliferation and apoptosis. Finally the differentiation, growth, and maturation of the axial skeletal elements may also play an important role (Table 3.3).

Pathogenesis of Abnormal Axial Development

The identification of the structural defect in congenital scoliosis in the fetus or neonate remains an analysis conducted long after the initial pathogenic events inducing the malformation. Identifying and understanding the initial pathogenic event is a critical step in characterizing the mechanisms of teratogenesis, which can

Table 3.3 Phases of somitogenesis in a stage 12 chick embryo and possible causal links between teratogen target tissues and hypothesized mature dysmorphogenesis

Stage 12 Chick Embryo	Transverse section/time of teratogenic insult	Target tissue	Possible resultant dysmorphogenes
	C	1. Notochord	Cleft vertebrae
		2. Ectoderm/ neural tube	Vertebral element agenesis
		3. Sclerotome	Vertebral disc anomalies
			Abnormal bone metabolism
	B	1. Notochord	Cleft vertebrae
		2. Ectoderm/ Neural tube	Vertebral agenesis
		3. Somitic mesoderm	Hemivertebrae Block vertebrae
		4. Lateral plate mesoderm	Bifid ribs
	A	1. Chordomesoderm/ notochord	Vertebral disk anomalies Caudal agenesis
		2. Paraxial mesoderm	Vertebral agenesis Hemiblock Vertebrae
		3. Ectoderm	Block or hemivertebrae Bifid or fused ribs vertebral agenesis

Phases of somitogenesis at three anterioposterior locations (A, B, and C) in a stage 12 chick embryo. Labels: DSo, differentiated somite; ESo, epithelial somite; CSo, condensed somite; PM, paraxial mesoderm; NT, neural tube; HN, Hensen's node; ECT, ectoderm; END, endoderm, NC, notochord; DM, dermomyotome; SC, sclerotome; IM, intermediate mesoderm; LM, lateral plate mesoderm.

then lead to the development of appropriate interventions. Environmental insults to a developing organism occur at molecular or sub-cellular levels. While the list of possible environmental insults is very large, the insults may be translated into types of cellular responses that result in recognizable patterns of dysfunction of dysmorphogenesis among tissues and organs (Table 3.4) (Wilson 1977, Sulik 1997).

Although teratogens are often discrete in nature (e.g., of known structure/composition and chemical characteristic), the determination of teratogenic mechanism is complicated. The main reasons for this are as follows. First, not all the possible targets of a teratogen have been identified, since many potentially affected targets remain unknown, i.e., normal developmental mechanisms still need identification and characterization. Related to this is the fact that it is highly unlikely that most teratogens act upon a single molecule or even a cellular pathway.

Table 3.4 Potential mechanisms, routes of pathogenesis, and ultimate morphogenetic outcome used by environmental teratogens in the induction of congenital malformation. Adapted from Wilson (1977)

Mechanisms	Pathogenesis	Final tissue outcome
Genetic mutation	Increased or decreased cell death	Reduction of cells to allow proper morphogenesis or tissue maturation
Chromosomal damage	Failed cell-cell interactions	
Epigenetic alteration	Reduced matrix biosynthesis	Imbalances in differentiation
Mitotic interference	Impeded morphogenic movements	Imbalances in growth
Nucleic acid synthesis/balance	Mechanical disruption of tissues	
Altered enzymatic substrates, co-factors, etc.		
Altered energy source		
Altered redox status		
Disrupted membrane or cytoskeletal integrity		
Altered signal transduction		

Multiple mechanistic pathways may combine to produce a single pathogenic mechanism contributing to the resultant congenital defect. Second, our ability to monitor the effect of the teratogen on the biochemistry of individual cellular targets is limited. Specifically, probes with sufficient sensitivity and specificity are unavailable for many processes and applications. Contributing to these issues is the fact that the amount of tissue available for study is usually very limited. Intertwined within these shortcomings is the difficulty of experimental interpretation, which varies with probe, detection, and particularly the endpoint chosen.

Despite these complications, we can hypothesize several intracellular processes that may be targets of teratogens. The teratogen generates its effects on the embryo often through a mode of molecular mimicry co-opting or undermining normal cellular processes such that they are activated, inactivated, or diverted in a manner inconsistent with developmental timing. Such processes include mitotic interference (mutagenesis and carcinogenesis), epigenetic changes (methylation and acetylation state), altered membrane function (composition or porosity), altered signal transduction, altered/inhibited energy metabolism, inhibition of waste (intermediary) metabolism, changed redox status, specific or general enzyme inhibition, and disturbances in nucleic acid synthesis, among other possibilities. The cellular responses to these insults may be grouped into several common outcomes, including necrosis or apoptosis, reduced biosynthesis, failed cell/cell or cell/matrix interactions, impeded morphogenetic movement, and mechanical disruption of tissues. Ultimately, the final defect may be manifested via loss of cells or tissue or imbalances in growth and differentiation.

While specific mechanisms of many teratogenic insults remain largely unknown, the characterization of cellular responses has been more successful. One particularly well-characterized outcome is the correlation of tissue-specific patterns of cell death and impending malformation (Sadler and Hunter 1994, Sulik et al. 1988, Sulik 1997, Zakeri and Ahuja 1997). This correlation has highlighted several characteristics of teratogenic action including the principles that different cell populations are sensitive to teratogenic insults at different time points, different agents target different tissues, and many teratogens expand areas of normal, developmentally regulated cell death. The observed changes in normal cell death patterns indicate the target tissue often plays a role in the subsequent dysmorphogenesis; however, the apoptotic cells do not participate in subsequent tissue formation – thus the effect of the teratogen on the surviving cells is important and presumed to be related to the cause of cell death. Nonetheless, the increase in cell death serves as an early marker for the teratogenic action. As we learn more about development and toxicological responses on the molecular level, we will create more sensitive cell response markers that will allow greater resolution of the teratological action.

Overview of Agents and Conditions Associated with Axial Skeletal Teratogenesis

As stated above, axial skeletal malformations are often linked to exposure to teratogenic conditions. The following summarizes the types of teratogens and teratogenic conditions associated with spinal malformations (Schardein 2000). Detailed descriptions of some of these factors will be presented in the following sections.

Recreational teratogens: Recreational drugs such as alcohol, cocaine, and cigarettes are known to significantly reduce fetal and post-natal growth, increase infant mortality, and cause congenital malformations of various types and severity.

Pharmaceutical teratogens: Most embryonic organs and the central nervous system are extremely sensitive to the teratogenic affects of pharmaceuticals such as thalidomide, diethylstilbestrol, retinoic acid, valproic acid, warfarin, chemotherapy, lithium, and nicotinic acid.

Industrial and environmental teratogens: Industrial processes required to provide for growing populations worldwide release a substantial amount of waste products into the environment, with the toxicologic and teratogenic effect of many species as yet uncharacterized. Among the chemicals with known teratogenic effects are organic solvents; arsenic, cadmium, and lead anesthetic gases; and organic mercury.

Agricultural teratogens: Insecticides and herbicides are critical to providing nutrition to growing populations. Studies have determined that organochlorine insecticides such as DTT, parathion, and malathion may interfere with fertility and reproduction by mimicking estrogen-like compounds. Among herbicides, the byproduct of Agent Orange, 2,3,7,8-tetrachloro-dibenzo-*p*-dioxin (TCDD), is highly teratogenic causing cleft palate and congenital renal abnormalities.

Infectious diseases: Microbial chemicals may act as teratogens. Microbes such as syphilis, cytomegalovirus, rubella, herpes, toxoplasma, and fifth disease affect 1–5% of all live births. These infections may cause a group of associated malformation known as the TORCH complex, as well as isolated structural defects and functional deficits.

Metabolic conditions: Some metabolic disorders, most prominently diabetes and hyperthermia also induce congenital malformations in the embryos. Diabetic pregnancy increases the frequency of a wide variety of congenital defects over background including cardiac defects, eye and ear defects, renal defects, and functional deficits in addition to a high rate of congenital scoliosis, in addition to increased embryonic death and life-long metabolic disorders.

Non-genetically Linked Conditions Characterized by Axial Skeletal Defects: VATER Association

The VATER spectrum is a non-random association characterized by vertebral anomalies (V), anal atresia (A), tracheoesophageal (TE) fistula, and renal (R) anomalies (Botto et al. 1997, Cohen 1997, Martínez-Frías and Frías 1997). This spectrum may also be associated with cardiovascular (C) anomalies and limb (L) anomalies (VACTERL). The incidence of VATER in diabetic mothers is 200× higher than in the general population, which occurs at a rate of 16 per 100,000 births (Pauli 1994, Cohen 1997, Martínez-Frías et al. 1998a). Vertebral defects in this association can involve agenesis, hypoplasia, and hemivertebrae, often afflicting many contiguous vertebral units. As the acronym suggests, many associated tissues are affected. The association of these different mesenchymally derived tissues to the vertebral column and the timing of their development are critical to hypothesizing the origin and mechanism of the defect(s). Analyses of the frequency and co-occurrence of the features of VACTERL and other syndromes suggest that the anomalies can extend to various cranio-caudal levels suggesting a time dependency and critical period through a defect in a common mechanism of dysmorphogenesis (Stewart et al. 1993). The VACTERL sequence can be conceptually included in a group of progressively severe spectrums of which it may be the most severe (Table 3.5). This broad spectrum of malformations has been coined the axial mesodermal dysplasia complex (AMDC) (Stewart et al. 1993). Some confounding features to any hypothesis are the broad range of defects sometimes involving tissues derived from all germ layers, its largely spontaneous occurrence, and low rate of subsequent inheritance.

There are two related, non-exclusive models currently employed to explain the etiology of AMDC suggesting that the collection of defects may arise from a single environmental insult at a time early in post-implantation development. In the first theory, the embryo at the time of early gastrulation is comprised of a single morphogenetic field, the primary developmental field (Optiz et al. 2002, Martinez-Frias et al. 1998b). At this time the embryo responds essentially as a single, homogeneous

Table 3.5 Common features of different associations within the axial mesodermal dysplasia complex spectrum

Malformation	VACTERL	VATER	OAV	PIV	PHS
Vertebral	X	X	X	X	
Imperforate anus	X	X		X	X
Craniofacial	X		X		
Tracheal–esophageal fistula	X	X			
Renal abnormalities	X	X			X
Limb anomalies	X			X	X

(OAV) Oculo-auriculo-vertebral dysplasia, (PIV) Polyoligodactyly-imperforate anus-vertebral anomalies syndrome, (PHS) Pallister-Hall syndrome.

entity. The primary effect of the insult at this time is to affect growth (proliferation) within the embryo, drastically affecting the existence and position of organizing centers and tissue morphogenesis throughout the embryo as this primary developmental field subdivides into secondary developmental fields that will give rise to the various organs and structures of the embryo including the axial skeleton (Martinez-Frias and Frias 1999). If the insult occurs at this time, there is necessarily a wide range of structures affected (polytopic defects) often of mesenchymal origin, but involving ectodermal and endodermal germ layers as well.

A second variant on the theory holds that the broad spectrum of defects reflects a common mechanistic cause in many tissues of a more heterogeneous entity comprised of multiple secondary developmental fields, such that different tissues of the embryo respond in specific manners to produce the wide spectrum of observed defects (Martínez-Frías and Frías 1997, Bohring et al. 1999). The defect then is thought to lie more in mechanisms of patterning or morphogenesis as the insult or defect occurs slightly later in development. This latter variation on developmental field defects and the etiology of multiple congenital anomalies such as VATER appears to more easily explain the wide spectrum of cranio-caudal positions of the defects and the wide degree of severity observed in several multiple congenital defect associations by allowing for a longer critical period. Both of these theories have been characterized theoretically and statistically to the range of defects observed in infants born to diabetic mothers, one of the most frequently recognized "causative" factors of the VATER spectrum (Martínez-Frías et al. 1998a). The high incidence of the VATER and other AMDC variants in diabetic mothers suggests an etiology that involves a fundamental metabolic imbalance in energy production or a dysfunction in a critical component of the embryonic stress response. Some investigators have suggested that the defects may arise from malformation or dysfunction of the notochord, which is critical to the establishment and maintenance of embryonic axes and the patterning and differentiation of many mesenchymal tissues (Gilbert-Barness et al. 2001). It has been suggested that notochord mutants such as brachyury (*T*) or sonic hedgehog (*SHH*) knock-outs could be used as models for VATER and AMDC (Arsic et al. 2002). We discuss the potential for energy metabolism dysregulation as the locus affected resulting in VATER in the context of diabetes-induced congenital scoliosis below.

Environmental Factors That Cause Axial Skeletal Dysmorphogenesis

Valproic Acid

Valproic acid (VPA) is an anti-epileptic drug that is associated with a 20-fold increased incidence of spina bifida, a neural tube defect, in children born to pregnant mothers undergoing VPA treatment (Lammer et al. 1987, Nau et al. 1991). Experimentally, VPA has been shown to be teratogenic in mouse, rat, chick, hamster, rabbit, and rhesus monkeys (Ehlers et al. 1992, Menegola et al. 1996, Vorhees et al. 1987, Barnes et al. 1996b, Basu and Wezeman 2000, Hendrickx et al. 1988). Skeletal abnormalities in these models were most commonly observed, involving vertebrae, ribs, digits, and craniofacial bones. These frequently occur in the context of other cardiovascular, urogenital, and neurological anomalies that together comprise the fetal valproate syndrome. The axial skeletal defects can include presacral vertebrae, cervical and thoracic ribs, indicating possible homeotic transformations. The defects may also include structural vertebral defects, indicating segmentation defects.

In general, the primary locus of teratogens causing spina bifida including VPA is believed to be the neural tube, resulting in failure of neural tube closure (Turner et al. 1990). Subsequently, the neural arches are unable to fuse. However, vertebral defects such as block vertebrae and hemivertebrae sometimes coincide with a neural tube defect have also been observed following VPA exposure (Barnes et al. 1996b). More detailed studies have shown that important patterning genes, such as *Pax1* and paraxis (*Tcf15*), are down-regulated by the administration of VPA in chick embryos (Barnes et al. 1996b, Barnes et al. 1997). The malformations produced by VPA can be mimicked through the administration of anti-sense deoxynucleotides during somitogenesis (Barnes et al. 1996b, Barnes et al. 1997). This type of data confirms that dysregulation of these genes can be teratogenic, but does not indicate a specific mechanism for how this may occur. The down-regulation of these genes may be caused by decreased signaling or reduced or delayed differentiation caused by increased ROS production or altered nucleic acid metabolism (Fantel 1996, Nau et al. 1991), as suggested by studies showing that folic acid administration can significantly reduce the incidence of experimentally induced VPA axial skeletal defects (Green and Copp 2005). More recently, VPA has been shown to also inhibit histone deacetylase activity at therapeutic levels, and that this activity is correlated with axial skeletal defects and exencephaly (Menegola et al. 2006). In a comparison of the teratogenicity and changes in gene expression by VPA and TSA (Trichostatin A), many of the shared genetic effects were specific to skeletal and cardiac muscle, assigning a more specific mechanistic action of VPA to dysregulating epigenetic control which leads to altered gene expression.

Hypoxia

Congenital vertebral anomalies have been produced in newborn animals experimentally by transient hypoxia and transient exposure during the embryonic period

(Grabowski and Paar 1958, Ingalls and Philbrook 1958, Rivard 1986, Webster and Abela 2007). In these studies, many gross vertebral and associated skeletal defects have been induced, including hemivertebrae, vertebral fusions, fragmented vertebral bodies, bifid ribs, or junctions of two or more ribs. The nature and extent of skeletal malformations induced have been dependent upon the precise stage of somite formation at the time when maternal stress has been induced. Hypoxia is thought to affect the early embryo through the induction of increased reactive oxygen species (ROS) homotopically (where ROS are already prominent; Fantel 1996) and later through altered vascularization (Grabowski 1961, Danielson et al. 1992, Webster and Abela 2007). Less well defined is the idea that hypoxia itself or its management is important in and of itself for morphogenic process or cell function during embryonic development (Chen et al. 1999, Semenza 1999).

During early organogenesis as the embryonic circulation develops, the embryo is known to undergo a transition from anaerobic respiration to aerobic respiration (Hunter and Sadler 1989, Hunter and Tugman 1995, Mackler et al. 1975, Miki et al. 1988a). Recent studies have confirmed that oxygenation and the cellular response to oxygenation as interpreted through expression patterns of heat shock proteins (protective chaparones) (Edwards et al. 1997, Mirkes 1997), antioxidant (superoxide dismutases) (Wells and Winn 1996, Ornoy 2007, Forsberg et al. 1996, Yon et al. 2008, Zaken et al. 2000), and HIF1alpha expression (Iyer et al. 1998, Maltepe et al. 1997, Jain et al. 1998, Minet et al. 2000) vary between different tissues of the embryo over time. Some of these variations have been correlated to periods of teratogenic susceptibility (Ornoy et al. 1999, Forsberg et al. 1996). In this transition, mitochondrial respiration may be inefficient producing higher-than-usual amounts of ROS at a time when embryonic defenses against ROS damage are not well developed. This combination can lead to excess ROS-induced cell stress and cell death (Dennery 2007, Dumollard et al. 2007, Burton et al. 2003). One hypothesis is that those tissues undergoing energetically demanding process such as morphogenesis are most susceptible to the oxygenation transition, a hypothesis furthered in diabetic embryopathy. The neural tube and somatic mesoderm have been shown to have a higher metabolic activity (Raddatz and Kucera 1983, Miki et al. 1988a, b, Mackler et al. 1971, Mackler et al. 1975) than surrounding tissues during early organogenesis, the time of greatest susceptibility to environmentally induced axial skeletal defects.

Carbon Monoxide

Early work studying the effects of hypoxia utilized carbon monoxide as a chemical hypoxic agent. Carbon monoxide (CO) is an odorless, colorless, non-irritating gas produced by the incomplete combustion of carbon containing materials. There have been no epidemiological studies of the direct effect of CO on human pregnancies (Schardein 2000). However, there are a number of case reports and anecdotes suggesting that CO may be a teratogen in humans (Robkin 1997, Longo 1977). Anecdotal accounts were given in Brander (Robkin 1997) and reported congenital malformations, such as microcephaly, micrognathia, and limb defects including hip dysplasia, tetraplegia, equinovarus, and limb reduction. Indirect epidemiological information can be obtained from the observations of pregnancy outcomes among women who smoke. Maternal smoking is associated with various adverse outcomes including low birth weight, decreases in successful births (Fichtner et al. 1990), and various behavioral defects that can be mimicked by CO alone in animal models (Bnait and Seller 1995).

There are a limited number of studies linking CO to congenital malformations. Early studies in chick, rabbit, and rat showed a causative relationship (Baker and Tumasonis 1972, Murray et al. 1979); however, later studies failed to confirm this connection (Astrup et al. 1975). More recent studies exploring threshold levels and critical periods related to CO-induced effects upon the embryo have documented CO-induced dysmorphogenesis (Bailey et al. 1994, Daughtrey et al. 1983, Loder et al. 2000, Alexander and Tuan 2003). CO exposures during early organogenesis, the critical period, resulted in vertebral anomalies, microphthalmia, and a phenotype similar to caudal dysgenesis syndrome. Such malformations have been reported with CO exposures administered during organogenesis in the context of other teratogens at sub-teratogenic levels (Singh et al. 1993, Singh 2006). This may be a significant problem worldwide since acute carbon monoxide exposures may be higher and more frequent than often reported (Fichtner et al. 1990, Ralston and Hampson 2000).

CO does impair oxygen delivery to and into cells by binding hemoglobin, myoglobin, and other porphyrins; however, it may also function as a signaling molecule in the context of nitric oxide (NO) signaling (Maines 1997). When administered after the vascular system is developed, the axial defects caused by CO are attributed to vascular leakage and subsequent mechanical disruption of developing tissue (Baker and Tumasonis 1972). However, during early organogenesis, axial defects involve the reduction of important segmentation genes including *Pax1* and paraxis (*Tcf15*) (Alexander and Tuan 2003), resulting possibly in the impaired inductive interaction of the neural tube with the paraxial mesoderm with CO acting as a signaling molecule. Nitric oxide is known to regulate neurulation and other early embryonic processes (Lee and Juchau, 1994), and CO can alter the production of NO in axial tissues (Alexander et al. 2007a). The impaired interaction is likely due to a loss of cell function characterized or indicated by increased neural tube apoptosis and loss of neural tube-derived somite epithelialization signals.

Diabetes

Maternal diabetes is known to have many teratogenic effects (Finnell and Dansky 1991, Aberg et al. 2001). Malformations including neural tube defects, caudal dysgenesis, vertebral defects, congenital heart defects, femoral hypoplasia, renal, and craniofacial anomalies are described in infants of diabetic mothers. Caudal regression syndrome is a severe condition characterized by agenesis, regression, and or disorganization of the posterior (sacral–lumbar) vertebrae and the malformation of the soft tissue at that level and below (Bohring et al. 1999, Martinez-Frias et al. 1998b). It occurs 200 times more frequently in diabetic than in nondiabetic pregnancies. Other major malformations of the midline are also much more frequent including VATER, OAV, and other major malformations. Together, these can be placed in a related and progressive spectrum of syndromes and non-random associations belonging to the ADMC.

Mouse models utilizing "diabetic environments" or hyperglycemia report various anomalies encompassing the full spectrum of embryonic embryopathy (Akazawa 1995, Ornoy et al. 1999). These models together reveal that hyperglycemia is sufficient to cause most of the defects observed in diabetic embryopathy including neural tube defects, axial skeletal defects, heart and craniofacial abnormalities, rib and renal defects – although no individual model phenocopies the condition completely. At physiological levels of hyperglycemia or ketosis, the most consistent outcome is a failure of anterior and posterior neuropore closure (Sadler et al. 1988, Sadler and Horton 1983, Horton and Sadler 1983, Ornoy et al. 1986, Sadler et al. 1989). Researchers have determined that the diabetic environment increases ROS production in these regions of the neural tube and in the primordia of the organs listed above including craniofacial region, otic, and optic cups, Hensen's node, and the notochord, caused by the diabetic environment. Coincident with the high ROS is an increase in cell death and a decrease in *Pax3*, a factor critical in neural tube closure (Fine et al. 1999, Loeken 2005). Application of folic acid and other antioxidants greatly reduced the incidence of ROS production (Ornoy 2007), insipient cell death and the reduction of *Pax3* expression.

The caudal agenesis/dysgenesis syndrome can be phenocopied by prolonged exposure to hyperglycemia, hyperketonemia, and streptozotocin. The collection of defects in these severely affected animals indicates an early patterning event is disturbed. The notochord is laid down during gastrulation and is responsible for dorsoventral and mediolateral patterning as well as survival of the mesoderm during axis elongation. High levels of cell death in the notochord are observed in severely affected animals, suggesting the notochord function is likely to be compromised, and mutations in the T-box gene brachyury (Rashbass et al. 1994) and disruption of *Shh* function (Kim et al. 2001) have been presented as possible models of caudal dysgenesis and other manifestations of ADMC.

The incidence and severity of malformations in diabetic pregnancies are correlated with poor glycemic control in the first trimester and can be reduced by instituting tight glycemic control prior to conception, and the evidence presented above of various antioxidants and insulin provides hope that a cocktail can be developed and delivered harmlessly to prevent the initiation of the diabetic embryopathic condition. While prevention of the condition appears at hand, the initial biochemical imbalance presents us with an interesting pattern. The condition of hyperglycemia provides a "free" energy source that is readily available to the mitochondrion for ATP production, a condition opposite to hypoxia, in which ATP production in greatly decreased. A reasonable hypothesis incorporating these two opposite conditions is that molecular regulation of any developmental process can be disturbed by abnormal maternal fuel metabolism, and the timing of specific episodes of poor glycemic control determines which organ systems are affected.

Retinoic Acid

Retinoic acid (RA) is an analog of vitamin A commonly used to treat acne and other skin conditions. In humans, prenatal exposure results in a characteristic pattern of defects including abnormalities in the ears, mandibles, palates, aortic arch, and central nervous system. In animal models, many similar defects are observed (Gudas 1994). At higher doses delivered during organogenesis, RA can induce axial skeletal defects as well as including homeotic transformations (Kessel and Gruss 1991, Rubin and LaMantia 1999, Kawanishi et al. 2003), and at higher doses axial skeletal truncations (Padmanabhan 1998).

Aside from being a well-characterized teratogen, retinoic is also a naturally occurring chemical involved in many aspects of embryonic patterning, including the patterning of the somites. The teratogenic effects of retinoic acid above are consistent with the in situ expression of RA receptors (Maden 1994, Cui et al. 2003, Iulianelle et al. 1999) and metabolic-transforming enzymes (Swindell et al. 1999, Niederreither et al. 2002, Reijntjes et al. 2004, Cammas et al. 2007) as well as the effect of knocking down these molecules in murine models. RA, its receptors, and *CYP26* are expressed in the paraxial mesoderm and act as critical regulators in the coordination of the somitogenesis clock and HOX gene expression (Duester 2007, reviewed in Sewell and Kusumi 2007). At higher levels, it is hypothesized that RA interrupts tissue morphogenesis, neural crest migration, and at highest doses causes cell death in morphogenically critical tissues, such as the neural tube, notochord, and paraxial mesoderm, resulting in a phenotype similar to caudal dysgenesis syndrome (Iulianelle et al. 1999).

Hyperthermia

Exposure of the human fetus to high temperatures (for example, 2°C over normal), as in the case of high fever or prolonged hot tub usage, is associated with neural tube defects, heart defects, microphthalmia, and functional deficits (Graham et al. 1998, Edwards et al. 1997). There is no epidemiological evidence suggesting heat shock causes axial skeletal defects. In studying the mechanisms of heat shock teratogenesis in animal models, vertebral defects were observed in many species including mice, rats, and chicks (Breen et al. 1999, Mirkes and Cornel 1992, Primmet et al. 1988). The severity of these defects is correlated to the time and duration of exposure. Experimental studies in chick embryos revealed that at moderate levels and exposure times (42°C for 20 minutes), one or two adjacent segments were fused into a single large somite. This effect was repeated every 7–8 somites separated by normal somitogenesis (Primmet et al. 1988). This result suggested a cell cycle-dependent mechanism to the defect and to somitogenesis itself, prompting the proposal of a clock and wave-front model for the patterning process of somitogenesis (Primmet et al. 1989).

The response of the embryo to hyperthermia is very dependent upon the degree of temperature increase, its duration, and the stage at which the heat shock is experienced (Graham et al. 1998). There is a steep threshold for embryonic survival and resorption, which suggests the general outcome of hyperthermia is embryonic resorption. At levels of hyperthermia inducing embryonic survival and malformation, tissue-specific cell death is observed. Investigators identified the induction of heat shock proteins (HSP) as a prominent feature of the embryonic response. These molecular chaperones play important roles in regulating protein folding during normal cell function, but they also serve to protect cells from environmental insult. In the process, the HSP-bound proteins are not able to perform their function (Buckiova et al. 1998, Walsh et al. 1999). During teratological doses of hyperthermia, the cell cycle is slowed, suggesting a mechanism of the vertebral anomalies observed. Recently the mechanism of somitogenesis was shown to involve the tightly controlled, cyclic expression of a variety of proteins many belonging to the Notch/Delta signaling system (Shifley and Cole 2007). During heat shock, some of these proteins or their targets may be bound by HSP, and we can hypothesize that this would disrupt the somitogenic clock resulting in disrupted pattern and ultimately vertebral defects. An important feature of the protective heat shock response then and its relation to teratogenesis is that their activation and function may reduce or delay tissue development or morphogenic actions. In fact, many teratogenic insults induce HSP activity, and as such HSP activation may be an underlying commonality in teratogenic mechanisms along with ROS production and apoptosis.

Arsenic

Arsenic, a metal pollutant, is found naturally in groundwater and unnaturally in mine waste sites, industrial byproducts, and in agricultural runoff. It is toxic to humans and is known to cause birth defects including spina bifida, craniofacial defects, developmental retardation and to decrease birth weight and increase incidences of fetal mortality, miscarriage, and still birth (Willhite and Ferm 1984, DeSesso et al. 1998). In experimental in vitro models, arsenic is teratogenic in mice, rats, and chicks (Hood and Bishop 1972, Chaineau et al. 1994, Beaudoin 1974, Lindgren et al. 1984, Peterková and Puzanová 1976, DeSesso et al. 1998), with neural tube defects being common among all of them (Shalat et al. 1996, Takeuchi 1979). Its toxicity is greatly dependent on its redox state: arsenate vs. arsenite. The structure of arsenate can mimic that of phosphate groups, imparting arsenate with the ability to disrupt various cell processes including nucleic acid metabolism, lipid metabolism, and electron transport. Inefficient electron transport can lead to high production of ROS, which have documented cell destructive activities and teratogenic capacity (Hunter 2000, Kitchin and Ahmad 2003, Bernstam and Nriagu 2000). In addition, arsenate can reduce to arsenite. The effects of arsenite on disruption of cell cycle and cytoskeletal structure have been attributed to its reaction to sulfhydryl groups (Levinson et al. 1980), which may account for its strong induction of the heat shock response (German et al. 1986, Mirkes and Cornel 1992,

Bernstam and Nriagu 2000). In addition, arsenite can disrupt the citric acid cycle and electron transport via binding to thiol group enzymatic active sites (DeSesso et al. 1998). A disruption in the energy status of different tissues of the developing embryo is attributed to teratogenicity of arsenic causing similar malformations to those observed in hypoxic or hyperglycemic environment; however, arsenic has other distinctive effects on the embryo (DeSesso et al. 1998). Arsenic and other metal compounds are very effective inducers of the heat shock response (Mirkes and Cornel 1992, Bernstam and Nriagu 2000), which may protect cell from molecular damage, but induce birth defects in its own right via disruption of the cell cycle and other cyclic and time-dependent morphogenetic processes (Wlodarczyk et al. 1996). In addition, cells surviving the initial arsenic insult, may pass on genetic damage that contribute to subsequent carcinogenic transformation later in ontogeny (Bernstam and Nriagu 2000). These multiple, interacting mechanisms may account for the wide range of malformations observed following acute arsenic exposure.

Ethanol

Ethanol, widely consumed as a recreational drug, has long been strongly associated with teratogenesis as fetal alcohol syndrome (FAS). FAS is present in one in three children of alcoholic mothers, with an estimated 40,000 children born every year in the United States (Schardein 2000, Thackray and Tift 2001). FAS manifestations include growth deficiency, central nervous system problems, characteristic facial features, and organ malformations. Features of FAS have been observed in animal models exposed to ethanol in utero or in vitro, including mice, rat, chick, and others (Sulik et al. 1981, Becker et al. 1996, Fernandez et al. 1983, Sanders and Cheung 1990, Yelin et al. 2007, Schardein 2000, Chaudhuri 2000, Padmanabhan and Muawad 1985).

The mouse model has been a particularly effective model in elucidating the etiology of ethanol-induced birth defects. One mechanism, of ethanol-induced teratogenesis is through ethanol impaired placental blood flow to the fetus by constricting blood vessels and inducing embryonic/fetal hypoxia and malnutrition (Shibley et al. 1999). Since ethanol rapidly crosses the placenta into the fetus, there are other direct embryonic and fetal targets of ethanol. The mouse model has been a particularly effective model in elucidating the etiology of ethanol-induced birth defects. Ethanol has been shown to increase cell death in critical cell populations including anterior neural folds and neural crest cells (Sulik et al. 1988, Rovasio and Battiato 1995, Dunty et al. 2001), which play a critical role in the morphogenesis of the face. Neural crest cells are particularly vulnerable to ethanol, inducing delayed/altered migration and cell death (Rovasio and Battiato 1995). Correlations have been made to increased ROS production within the neural crest population (Kotch et al. 1995), mitochondrial dysfunction and cell death in the etiology of ethanol and other teratogens (Ornoy 2007). The anterior neural tube and cranial neural crest have been the subject of intense scrutiny in the teratogenic mechanisms of ethanol, however, other tissues are also affected, including the eye, ear, heart, renal system, and axial skeleton (Kennedy and Elliott 1986, Parnell et al. 2006, Sulik 2005, Webster and Abela 1984, Assadi and Zajak 1992, Sanders and Cheung 1990, Carvan et al. 2004). With respect to the axial skeleton, investigators observed a misalignment or segmentation defect in ethanol-exposed embryos. Despite the substantial morphological difference with heat shock- treated embryos, the investigators suggested that the mechanism may be similar to heat shock (Carvan et al. 2004), involving the induction of the stress response by increased ROS production.

Methanol

Methanol is an alcohol encountered frequently during industrial processes. When the effects of inhaled ethanol and methanol were compared, the highest doses of methanol significantly increased the incidence of various defects including skeletal malformations (Nelson et al. 1985). Skeletal malformations were the most prevalent congenital defects observed and included vertebral abnormalities and an increased incidence of cervical ribs. Other skeletal abnormalities caused by methanol have been observed including holoprosencephaly, facial dysmorphogenesis, basicranial malformation, duplications of the atlas and axis and cervical vertebral abnormalities, and abnormal number of presacral vertebrae (Connelly and Rogers 1997, Rogers et al. 2004). Initial cellular responses appear similar to ethanol at the level of tissue-specific cell death (Abbott et al. 1995). In contrast to ethanol, many of the axial skeletal defects indicate homeotic shifts in segment identity.

Conclusion

Advances in cell and molecular biology with respect to normal development and somitogenesis and the pathogenesis and mechanisms of teratogenesis are occurring at a tremendous rate. This allows teratologists and developmental toxicologists the opportunity to revisit old problems with new tools. Despite the large number of cellular processes that may be disturbed by a teratogen, there are only a limited number of cellular and morphological outcomes. This has led investigators to strive for the identification of very defined critical periods and doses in a variety of model systems to aid in the identification of the initial targets of a teratogen and the true, hypothetically singular target molecule or process, as proposed by Wilson in 1956. Applying genomic and proteomic technologies to the problem of teratogenesis should begin to reveal the full spectrum of cellular processes affected, and elucidate links between variations in genotype and the effect of the environment on the phenotype that produce birth defects such as congenital scoliosis. The identification, at least in part, of this "holy grail" will aid in the development of new preventative treatments to a variety of teratogenic insults.

Acknowledgment Supported in by the Intramural Research Program of the National Institute of Arthritis, and Musculoskeletal and Skin Diseases, NIH (ZO1 AR41131).

References

Abbott, B.D., Ebron-McCoy, M., and Andrews, J.E. 1995. Cell death in rat and mouse embryos exposed to methanol in whole embryo culture. Toxicology 97:159–171.

Aberg, A., Westbom, L., and Källén, B. 2001. Congenital malformations among infants whose mothers had gestational diabetes or preexisting diabetes. Early Hum. Dev. 61:85–95.

Akazawa, S. 2005. Diabetic embryopathy: studies using a rat embryo culture system and an animal model. Cong. Anom. 45:73–79.

Alexander, P.G., Boyce, A.T., and Tuan, R.S. 2007a. Skeletal development. In Principles of Developmental Genetics, ed. S.A. Moody, pp. 866–905. New York, NY: Elsevier Academic Press

Alexander, P.G., Chau, L., and Tuan, R.S. 2007b. Role of nitric oxide in chick embryonic organogenesis and dysmorphogenesis. Birth Defects Res. A Clin. Mol. Teratol. 79:581–594.

Alexander, P.G. and Tuan, R.S. 2003. Carbon monoxide-induced axial skeletal dysmorphogenesis in the chick embryo. Birth Defect Res. Part A Clin. Mol. Teratol. 67:219–230.

Arsic, D., Qi, B.Q., and Beasley, S.W. 2002. Hedgehog in the human: a possible explanation for the VATER association. J. Paediatr. Child Health 38:117–121.

Assadi, F.K. and Zajac, C.S. 1992. Ultrastructural changes in the rat kidney following fetal exposure to ethanol. Alcohol 9:509–512.

Astrup, P., Trolle, D., Olsen, H.M. et al. 1975. Moderate hypoxia exposure and fetal development. Arch. Environ. Health 30:15–16.

Bailey, L.T., Johnston, M.C., and Billet, J. 1994. Effects of carbon monoxide and hypoxia on cleft lip in A/J mice. Cleft Palate-Craniofac. J. 32:14–19.

Baker, F.D. and Tumasonis, C.F. 1972. Carbon monoxide and avian embryogenesis. Arch. Environ. Health 24:53–61.

Barnes, G.L., Alexander, P.G., Hsu, C.W. et al. 1997. Cloning and characterization if chicken *Paraxis*: a regulator of paraxial mesoderm development and somite formation. Dev. Biol. 189:95–111.

Barnes, G.L., Hsu, C.W., Mariani, B.D. et al. 1996a. Chicken *Pax-1* gene: structure and expression during embryonic somite development. Differentiation 61:13–23.

Barnes, G.L., Mariani, B.D., and Tuan, R.S. 1996b. Valproic acid induced somite teratogenesis in the chick embryo: relationship with *Pax-1* gene expression. Teratology 54:93–102.

Basu, A. and Wezeman, F.H. 2000. Developmental toxicity of valproic acid during embryonic chick vertebral chondrogenesis. Spine 25:2158–2164.

Beaudoin, A.R. 1974. Teratogenicity of sodium arsenate in rats. Teratology 10:153–157.

Becker, H.C., Diaz-Granados, J.L., and Randall, C.L. 1996. Teratogenic actions of ethanol in the mouse: a minireview. Pharmacol. Biochem. Behav. 55:501–513.

Bernstam, L. and Nriagu, J. 2000. Molecular aspects of arsenic stress. J. Toxicol. Environ. Health B Crit. Rev. 3:293–322.

Bnait, K.S. and Seller, M.J. 1995. Ultrastructural changes in 9-day old mouse embryos following maternal tobacco smoke inhalation. Exp. Toxicol. Pathol. 47:453–461.

Bohring, A., Lewin, S.O., Reynolds, J.F. et al. 1999. Polytopic anomalies with agenesis of the lower vertebral column. Am. J. Med. Genet. 87:99–114.

Botto, L.D., Khoury, M.J., Mastroiacovo, P. et al. 1997. The spectrum of congenital anomalies of the VATER association: an international study. Am. J. Med. Genet. 71:8–15.

Breen, J.G., Claggett, T.W., Kimmel, G.L. et al. 1999. Heat shock during rat embryo development in vitro results in decreased mitosis and abundant cell death. Reprod. Toxicol. 13:31–39.

Brent, R.L. and Fawcett, L.B. 2007. Developmental toxicology, drugs, and fetal teratogenesis. In Clinical Obstetrics: The Tetus and Mother, 3rd Ed., eds. E.A. Reece and J.C. Hobbins, pp. 217–235. Malden, MA: Blackwell Publishing, Inc.

Buckiova, D., Kubinova, L., Soukup, A. et al. 1998. Hyperthermia in the chick embryo: HSP and possible mechanisms of developmental defects. Int. J. Dev. Biol. 42:737–740.

Burton, G.J., Hempstock, J., and Jauniaux, E. 2003. Oxygen, early embryonic metabolism and free radical-mediated embryopathies. Reprod. Biomed. Online 6:84–96.

Cammas, L., Romand, R., Fraulob, V. et al. 2007. Expression of the murine retinol dehydrogenase 10 (Rdh10) gene correlates with many sites of retinoid signalling during embryogenesis and organ differentiation. Dev. Dyn. 236:2899–2908.

Carvan, M.J. 3rd, Loucks, E., Weber, D.N. et al. 2004. Ethanol effects on the developing zebrafish: neurobehavior and skeletal morphogenesis. Neurotoxicol. Teratol. 26:757–768.

Chaineau, E., Binet, S., Pol, D. et al. 1994. Embryotoxic effects of sodium arsenite and sodium arsenate on mouse embryos in culture. Teratology 41:105–112.

Chaudhuri, J.D. 2000. Alcohol and the developing fetus–a review. Med. Sci. Monit. 6:1031–1041.

Chen, E.Y., Fujinaga, M., and Giaccia, A.J. 1999. Hypoxic microenvironment within an embryo induces apoptosis and is essential for proper morphological development. Teratology 60: 215–225.

Christ B., Huang R., Scaal M. (2004) Formation and differentiation of the avian sclerotome. Anat Embryol 208:333–350.

Christ, B., Huang, R., Wilting, J. 2000. The development of the avian vertebral column. Anat. Embryol. 202:179–194.

Cohen, M.M. 1997. The Child with Multiple Birth Defects. pp. 3–14. New York, NY: Oxford University Press

Connelly, L.E. and Rogers, J.M. 1997. Methanol causes posteriorization of cervical vertebrae in mice. Teratology 55:138–144.

Cui, J., Michaille, J.J., Jiang, W. et al. 2003. Retinoid receptors and vitamin A deficiency: differential patterns of transcription during early avian development and the rapid induction of RARs by retinoic acid. Dev. Biol. 260:496–511.

Danielson, M.K., Danielsson, B.R., Marchner, H. et al. 1992. Histopathological and hemodynamic studies supporting hypoxia and vascular disruption as explanation to phenytoin teratogenicity. Teratology 46:485–497.

Daughtrey, W.C., Newby-Schmidt, M.B., and Norton, S. 1983. Forebrain damage in chick embryos exposed to carbon monoxide. Teratology 28:83–89.

Dennery P.A. 2007. Effects of oxidative stress on embryonic development. Birth Defects Res. C Embryo Today 81:155–162.

DeSesso, J.M., Jacobson, C.F., Scialli, A.R. et al. 1998. An assessment of the developmental toxicity of inorganic arsenic. Reprod. Toxicol. 12:385–433.

Dias, M.S. 2007. Normal and abnormal development of the spine. Neurosurg. Clin. N. Am. 18:415–429.

Duester, G. 2007. Retinoic acid regulation of the somitogenesis clock. Birth Defects Res. Part C 81:84–92.

Dumollard, R., Duchen, M., and Carroll, J. 2007. The role of mitochondrial function in the oocyte and embryo. Curr. Top. Dev. Biol. 77:21–49.

Dunty, W.C. Jr., Chen, S.Y., Zucker, R.M. et al. 2001. Selective vulnerability of embryonic cell populations to ethanol-induced apoptosis: implications for alcohol-related birth defects and neurodevelopmental disorder. Alcohol Clin. Exp. Res. 25:1523–1535.

Edwards, M.J., Walsh, D.A., Li, Z. 1997. Hyperthermia, teratogenesis and the heat shock response in mammalian embryos in culture. Int. J. Dev. Biol. 41:345–358.

Ehlers, K., Stürje, H., Merker, H.J. et al. 1992. Valproic acid-induced spina bifida: a mouse model. Teratology 45:145–154.

Erol, B., Kusumi, K., Lou, J. et al. 2002. Etiology of congenital scoliosis. U. Penn. Orthoped. J. 15:37–42.

Fantel, A.G. 1996. Reactive oxygen species in developmental toxicity: review and hypotheses. Teratology 53:196–217.

Fernandez, K., Caul, W.F., Boyd, J.E. et al. 1983. Malformations and growth of rat fetuses exposed to brief periods of alcohol in utero. Teratog. Carcinog. Mutagen. 3:457–460.

Fichtner, R.R., Sullivan, K.M., Zyrkowski, C.L. et al. 1990. Racial/Ethnic differences in smoking, other risk factors and low birth weight among low-income pregnant women, 1978–1988. M.M.W.R. C.D.C. Surveill. Summ. 39(3):13–31.

Fine, E.L., Horal, M., Chang, T.I. et al. 1999. Evidence that elevated glucose causes altered gene expression, apoptosis, and neural tube defects in a mouse model of diabetic pregnancy. Diabetes 48:2454–2462.

Finnell, R.H. and Dansky, L.V. 1991. Parental epilepsy, anticonvulsant drugs, and reproductive outcome: epidemiological and experimental findings spanning three decades; 1: Animal Studies. Reprod. Toxicol. 5:281–299.

Forsberg, H., Borg, L.A., Cagliero, E. et al. 1996. Altered levels of scavenging enzymes in embryos subjected to a diabetic environment. Free Radic. Res. 24:451–459.

German, J., Louie, E, and Banerjee, D. 1986. The heat-shock response in vivo: experimental induction during mammalian organogenesis. Teratog. Carcinog. Mutagen. 6:555–562.

Gilbert-Barness, E., Debich-Spicer, D. et al. 2001. Evidence for the "midline" hypothesis in associated defects of laterality formation and multiple midline anomalies. Am. J. Med. Genet. 101:382–387.

Grabowski, C.T. 1961. A quantitative study of the lethal and teratogenic effects of hypoxia on the three-day chick embryo. Am. J. Anat. 109:25–35.

Grabowski, C.T. and Paar, J.A. 1958. The teratogenic effects of graded doses of hypoxia on the chick embryo. Am. J. Physiol. 103:313–347.

Graham, J.M. Jr., Edwards, M.J., and Edwards, M.J. 1998. Teratogen update: gestational effects of maternal hyperthermia due to febrile illnesses and resultant patterns of defects in humans. Teratology 58:209–221.

Green, N.D.E. and Copp, A.J. 2005. Mouse models of neural tube defects: investigating preventative mechanisms. Am. J. Med. Genet. Part C 135:31–41.

Gridley, T. 2006. The long and short of it: somite formation in mice. Dev. Dyn. 235:2330–2336.

Gudas, L.J. 1994. Retinoids and vertebrate development. J. Biol. Chem. 269:15399–153402.

Healy, C., Uwanogho, D., and Sharpe, P.T. 1999. Regulation and role of Sox9 in cartilage formation. Dev. Dyn. 215:69–78.

Hendrickx, A.G., Nau, H., Binkerd, P. et al. 1988. Valproic acid developmental toxicity and pharmacokinetics in the rhesus monkey: an interspecies comparison. Teratology 38:329–345.

Hood, R.D. and Bishop, S.L. 1972. Teratogenic effects of sodium arsenate in mice. Arch. Environ. Health 24:62–65.

Horton, W.E. Jr. Sadler, T.W. 1983. Effects of maternal diabetes on early embryogenesis. Alterations in morphogenesis produced by the ketone body, B-hydroxybutyrate. Diabetes 32:610–616.

Hunter, E.S. 3rd. 2000. Role of oxidative damage in arsenic-induced teratogenesis. Teratology 62:240.

Hunter, E.S. 3rd and Sadler, T.W. 1989. Fuel-mediated teratogenesis: biochemical effects of hypoglycemia during neurulation in mouse embryos in vitro. Am. J. Physiol. 257:E269–276.

Hunter, E.S. 3rd and Tugman, J.A. 1995. Inhibitors of glycolytic metabolism affect neurulation-staged mouse conceptuses in vitro. Teratology 52:317–323.

Ingalls, T.H. and Philbrook, F.R. 1958. Monstrosities induced by hypoxia. New Eng. J. Med. 259:558–564.

Iulianella, A., Beckett, B., Petkovich, M. et al. 1999. A molecular basis for retinoic acid-induced axial truncation. Dev. Biol. 205:33–48.

Iyer, N.V., Kotch, L.E., Agani, F. et al. 1998. Cellular and developmental control of O2 homeostasis by hypoxia-inducible factor 1 alpha. Genes Dev. 12:149–162.

Jain, S., Maltepe, E., Lu, M.M. et al. 1998. Expression of ARNT, ARNT2, HIF1 alpha, HIF2 alpha and Ah receptor mRNAs in the developing mouse. Mech. Dev. 73:117–123.

Jaskwhich, D., Ali, R.M. et al. 2000. Congenital scoliosis. Curr. Opin. Pediatr. 12:61–66.

Kawanishi, C.Y., Hartig, P., Bobseine, K.L. et al. 2003. Axial skeletal and hox expression domain alterations induced by retinoic acid, valproic acid, and bromoxynil during murine development. J. Biochem. Mol. Toxicol. 17:346–356.

Kennedy, L.A. and Elliott, M.J. 1986. Ocular changes in the mouse embryo following acute maternal ethanol intoxication. Int. J. Dev. Neurosci. 4:311–317.

Kessel, M. and Gruss, P. 1991. Homeotic transformations of murine vertebrae and concomitant alteration of Hox codes induced by retinoic acid. Cell 67:89–104.

Kim, P.C.W., Mo, R., Hui, C.-C. 2001. Murine models of VACTERL syndrome: role of sonic hedgehog signaling pathway. J. Pediatr. Surg. 36:381–384.

Kitchin, K.T. and Ahmad, S. 2003. Oxidative stress as a possible mode of action for arsenic carcinogenesis. Toxicol. Lett. 137:3–13.

Kotch, L.E., Chen, S.Y., and Sulik, K.K. 1995. Ethanol-induced teratogenesis: free radical damage as a possible mechanism. Teratology 52:128–136.

Lammer, E.J., Sever, L.E., and Oakley, G.P. 1987. Teratogen up date: Valproic acid. Teratology 35:465–473.

Lee, Q.P. and Juchau, M.R. 1994. Dysmorphogenic effects of nitric oxide (NO) and NO-synthase inhibition: studies with intra-amniotic injections of sodium nitroprusside and NG-monomethyl-L-arginine. Teratology 49:452–464.

Levinson, W., Oppermann, H., Jackson, J. 1980. Transition series metals and sulfhydryl reagents induce the synthesis of four proteins in eukaryotic cells. Biochim. Biophys. Acta. 606: 170–180.

Lindgren, A., Danielsson, B.R., Dencker, L. et al. 1984. Embryotoxicity of arsenite and arsenate: distribution in pregnant mice and monkeys and effects on embryonic cells in vitro. Acta Pharmacol. Toxicol. (Copenh.) 54:311–320.

Loder, R.T., Hernandez, M.J., Lerner, A.L. et al. 2000. The induction of congenital spinal deformities in mice by maternal carbon monoxide exposure. J. Pediatr. Orthoped. 20:662–666.

Loeken, M.R. 2005. Current perspectives on the causes of neural tube defects resulting from diabetic pregnancy. Am. J. Med. Genet. C Semin. Med. Genet. C 135:77–87.

Longo, L.D. 1977. The biological effects of carbon monoxide on the pregnant woman, fetus, and the newborn infant. Am. J. Obstetr. Gynecol. 129:69–103.

Mackler, B., Grace, R., Duncan, H.M. 1971. Studies of mitochondrial development during embryogenesis in the rat. Arch. Biochem. Biophys. 144:603–610.

Mackler, B., Grace, R., Tippit, D.F. et al. 1975. Studies of the development of congenital anomalies in rats. III. Effects of inhibition of mitochondrial energy systems on embryonic development. Teratology 12:291–296.

Maden, M. 1994. Distribution of cellular retinoic acid-binding proteins I and II in the chick embryo and their relationship to teratogenesis. Teratology 50(4):294–301.

Maines, M.D. 1997. The heme oxygenase system: a regulator of second messenger gases. Ann. Rev. Pharmacol. Toxicol. 37:517–554.

Maltepe, E., Schmidt, J.V., Baunoch, D. et al. 1997. Abnormal angiogenesis and responses to glucose and oxygen deprivation in mice lacking the protein ARNT. Nature 386:403–407.

Martínez-Frías, M.L., Bermejo, E., Rodríguez-Pinilla, E. et al. 1998a. Epidemiological analysis of outcomes of pregnancy in gestational diabetic mothers. Am. J. Med. Genet. 78:140–145.

Martínez-Frías, M.L., and Frías, J.L. 1997. Primary developmental field. III: Clinical and epidemiological study of blastogenetic anomalies and their relationship to different MCA patterns. Am. J. Med. Genet. 70:11–15.

Martinez-Frias, M.L., and Frias, J.L. 1999. VACTERL as primary, polytopic, developmental field defects. Am. J. Med. Genet. 83:13–16.

Martinez-Frias, M.L., Frias, J.L., and Opitz, J.M. 1998b. Errors of morphogenesis and developmental field theory. Am. J. Med. Genet. 76:291–296.

Menegola, E., Broccia, M.L., Nau, H. et al. 1996. Teratogenic effects of sodium valproate in mice and rats at midgestation and at term. Teratog. Carcinog. Mutagen. 16:97–108.

Menegola, E., Di Renzo, F., Broccia, M.L. et al. 2006. Inhibition of histone deacetylase as a new mechanism of teratogenesis. Birth Defects Res. C Embryo Today 78:345–353.

Miki, A., Fujimoto, E., Ohsaki, T. et al. 1988a. Effects of oxygen concentration on embryonic development in rats: a light and electron microscopic study using whole embryo culture techniques. Anat. Embryol. 178:337–343.

Miki, A., Mizoguchi, A., Mizoguti, H. 1988b. Histochemical studies of enzymes of the energy metabolism in post implantation rat embryos. Histochemistry 88:489–495.

Minet, E., Michel, G., Remacle, J. et al. 2000. Role of HIF-1 as a transcription factor involved in embryonic development, cancer progression, and apoptosis. Int. J. Mol. Med. 5: 253–259.

Mirkes, P.E. 1997. Molecular/cellular biology of the heat stress response and its role in agent-induced teratogenesis. Mutat. Res. 396:163–173.

Mirkes, P.E. and Cornel, L. 1992. A comparison of sodium arsenite- and hyperthermia-induced stress responses and abnormal development in cultured postimplantation rat embryos. Teratology 46:251–259.

Murray, F.J., Schwetz, B.A., Crawford, A.A. et al. 1979. Embryotoxicity of inhaled sulfur dioxide and carbon monoxide in mice and rabbits. J. Environ. Sci. Health C 13:233–250.

Nau, H., Hauck, R.S., and Ehlers, K. 1991. Valproic acid-induced neural tube defects in mouse and human: aspects of chirality, alternative drug development, pharmacokinetics and possible mechanisms. Pharmacol. Toxicol. 69:310–321.

Nelson, B.K., Brightwell, W.S., MacKenzie, D.R. et al. 1985. Teratological assessment of methanol and ethanol at high inhalation levels in rats. Fundam. Appl. Toxicol. 5:727–736.

Niederreither, K., Fraulob, V., Garnier, J.M. et al. 2002. Differential expression of retinoic acid-synthesizing (RALDH) enzymes during fetal development and organ differentiation in the mouse. Mech. Dev. 110:165–171.

Nishimura, H., Tanimura, T., Semba, R. et al. 1974. Normal development of early human embryos: observation of 90 specimens at Carnegie stages 7 to 13. Teratology 10:1–5.

Opitz, J.M., Zanni, G., Reynolds, J.F. Jr. et al. 2002. Defects of blastogenesis. Am. J. Med. Genet. 115:269–286.

O'Rahilley, R., and Müller, F. 1996. Human Embryology and Teratology, 3rd Ed. New York, NY: Wiley-Liss Publishers

Ornoy, A. 2007. Embryonic oxidative stress as a mechanism of teratogenesis with special emphasis on diabetic embryopathy. Reprod. Toxicol. 24:31–41.

Ornoy, A., Zaken, V., and Kohen, R. 1999. Role of reactive oxygen species (ROS) in the diabetes-induced anomalies in rat embryos in vitro: reduction in antioxidant enzymes and low-molecular-weight antioxidants (LMWA) may be the causative factor for increased anomalies. Teratology 60:376–386

Ornoy, A., Zusman, I., Cohen, A.M. et al. 1986. Effects of sera from Cohen, genetically determined diabetic rats, streptozotocin diabetic rats and sucrose fed rats on in vitro development of early somite rat embryos. Diabetes Res. 3:43–51.

Oskouian, R.J., Sansur, C.A., and Shaffrey, C.I. 2007. Congenital abnormalities of the thoracic and lumbar spine. Neurosurg. Clin. N. Am. 18:479–498.

Padmanabhan, R. 1998. Retinoic acid-induced caudal regression syndrome in the mouse fetus. Reprod. Toxicol.12:139–511.

Padmanabhan, R. and Muawad, W.M. 1985. Exencephaly and axial skeletal dysmorphogenesis induced by acute doses of ethanol in mouse fetuses. Drug Alcohol Depend. 16:215–227.

Parnell, S.E., Dehart, D.B., Wills, T.A. et al. 2006. Maternal oral intake mouse model for fetal alcohol spectrum disorders: ocular defects as a measure of effect. Alcohol Clin. Exp. Res. 30:1791–1798.

Pauli, R.M. 1994. Lower mesodermal defects: a common cause of fetal and early neonatal death. Am. J. Med. Genet. 50:154–172.

Peterková, R., and Puzanová, L. 1976. Effect of trivalent and pentavalent arsenic on early developmental stages of the chick embryo. Folia Morphol. 24:5–13.

Primmett, D.R.N., Norris, W., Carlson, G. et al. 1989. Periodic segmental anomalies induced by heat-shock in the chick embryo are due to interference with the cell division cycle. Development 105:119–130.

Primmett, D.R.N., Stern, C.D., Keynes, R.J. 1988. Heat-shock causes repeated segmental anomalies in the chick embryo. Development 104:331–339.

Raddatz, E. and Kucera, P. 1983. Mapping of the oxygen consumption in the gastrulating chick embryo. Resp. Physiol. 51:153–166.

Ralston, J.D. and Hampson, N.B. 2000. Incidence of severe unintentional carbon monoxide poisoning differs across racial/ethnic categories. Public Health Rep. 115:46–51.

Rashbass, P., Wilson, V., Rosen, B. et al. 1994. Alterations in gene expression during mesoderm formation and axial patterning in Brachyury (T) embryos. Int. J. Dev. Biol. 38: 35–44.

Reijntjes, S., Gale, E., and Maden, M. 2004. Generating gradients of retinoic acid in the chick embryo: Cyp26C1 expression and a comparative analysis of the Cyp26 enzymes. Dev. Dyn. 230:509–517.

Rivard, C.H. 1986. Effects of hypoxia on the embryogenesis of congenital vertebral malformations in the mouse. Clin. Orthop. Relat. Res. 208:126–130.

Robkin, M.A. 1997. Carbon monoxide and the embryo. Int. J. Dev. Biol. 41:283–289.

Rogers, J.M., Brannen, K.C., Barbee, B.D. et al. 2004. Methanol exposure during gastrulation causes holoprosencephaly, facial dysgenesis, and cervical vertebral malformations in C57BL/6 J mice. Birth Defects Res. B Dev. Reprod. Toxicol. 71:80–88.

Rovasio, R.A. and Battiato, N.L. 1995. Role of early migratory neural crest cells in developmental anomalies induced by ethanol. Int. J. Dev. Biol. 39:421–422.

Rubin, W.W. and LaMantia, A.S. 1999. Age-dependent retinoic acid regulation of gene expression distinguishes the cervical, thoracic, lumbar, and sacral spinal cord regions during development. Dev. Neurosci. 21:113–125.

Sadler, T.W. and Horton, W.E. Jr. 1983. Effects of maternal diabetes on early embryogenesis. The role of insulin and insulin therapy. Diabetes 32:1070–1074.

Sadler, T.W., Hunter, E.S. 3rd, Balkan, W. et al. 1988. Effects of maternal diabetes on embryogenesis. Am. J. Perinatol. 5:319–326.

Sadler, T.W. and Hunter, E.S. III. 1994. Principles of abnormal development. In Developmental Toxicology, 2nd Ed., eds. C.A. Kimmel and J. Buelke-Sam, pp. 53–63. New York, NY: Raven Press, Ltd.

Sadler, T.W., Hunter, E.S. 3rd, Wynn, R.E. et al. 1989. Evidence for multifactorial origin of diabetes-induced embryopathies. Diabetes 38:70–74.

Sanders, E.J. and Cheung, E. 1990. Ethanol treatment induces a delayed segmentation anomaly in the chick embryo. Teratology 41:289–297.

Schardein, J.L. 2000. Chemically Induced Birth Defects, 3rd Ed., pp. 998–1000. New York, NY: Marcel Dekker, Inc.

Semenza, G.L. 1999. Perspectives on oxygen sensing. Cell 98:281–284.

Sewell, W. and Kusumi, K. 2007. Genetic analysis of molecular oscillators in mammalian somitogenesis: clues for studies of human vertebral disorders. Birth Defects Res. C Embryo Today 81:111–120.

Shalat, S.L., Walker, D.B., and Finnell, R.H. 1996. Role of arsenic as a reproductive toxin with particular attention to neural tube defects. J. Toxicol. Environ. Health 48:253–272.

Shibley, I.A. Jr., McIntyre, T.A., and Pennington, S.N. 1999. Experimental models used to measure direct and indirect ethanol teratogenicity. Alcohol 34:125–140.

Shifley, E.T. and Cole, S.E. 2007. The vertebrate segmentation clock and its role in skeletal birth defects. Birth Defects Res. C Embryo Today 81:121–133.

Singh, J. 2006. Interaction of maternal protein and carbon monoxide on pup mortality in mice: implications for global infant mortality. Birth Defects Res. B Dev. Reprod. Toxicol. 77: 216–226.

Singh, J., Aggison, L. Jr., and Moore-Cheatum, L. 1993. Teratogenicity and developmental toxicity of carbon monoxide in protein-deficient mice. Teratology 48:149–159.

Stewart, F.J., Nevin, N.C., and Brown, S. 1993. Axial mesodermal dysplasia spectrum. Am. J. Med. Genet. 45:426–429.

Stockdale, F.E., Nikovitis, W. Jr., Christ, B. 2000. Molecular and cellular biology of avian somite development. Dev. Dyn. 219:304–321.

Sulik, K.K. 1997. Pathogenesis of abnormal development. In Handbook of Developmental Toxicology, ed. R.D. Hood, pp. 43–60. New York, NY: CRC Press

Sulik, K.K. 2005. Genesis of alcohol-induced craniofacial dysmorphism. Exp. Biol. Med. 230:366–375.

Sulik, K.K., Cook, C.S., and Webster, W.S. 1988. Teratogens and craniofacial malformations: relationships to cell death. Development 103(Suppl):213–231.

Sulik, K.K., Johnston, M.C., and Webb, M.A. 1981. Fetal alcohol syndrome: embryogenesis in a mouse model. Science 214:936–938.

Swindell, E.C., Thaller, C., Sockanathan, S. et al. 1999. Complementary domains of retinoic acid production and degradation in the early chick embryo. Dev. Biol. 216:282–296.

Tam, P.P. and Trainor, P.A. 1994. Specification and segmentation of the paraxial mesoderm. Anat. Embryol. 189:275–305.

Takeuchi, I.K. 1979. Embryotoxicity of arsenic acid: light and electron microscopy of its effect on neurulation-stage rat embryo. J. Toxicol. Sci. 4:405–416.

Thackray, H. and Tifft, C. 2001. Fetal alcohol syndrome. Pediatr. Rev. 22:47–55.

Turner, S., Sucheston, M.E., De Philip, R.M. et al. 1990. Teratogenic effects on the neuroepithelium of the CD-1 mouse embryo exposed in utero to sodium valproate. Teratology 41:421–442.

Vorhees, C.V. 1987. Teratogenicity and developmental toxicity of valproic acid in rats. Teratology 35:195–202.

Wallin, J., Wilting, J., Koseki, H. et al. 1994. The role of Pax-1 in axial skeleton development. Development 120:1109–1121.

Walsh, D., Grantham, J., Zhu, X.O. et al. 1999. The role of heat shock proteins in mammalian differentiation and development. Environ. Med. 43:79–87.

Webster, W.S. and Abela, D. 2007. The effect of hypoxia in development. Birth Defects Res. C Embryo Today 81:215–228.

Wells, P.G. and Winn, L.M. 1996. Biochemical toxicology of chemical teratogenesis. Crit. Rev. Biochem. Mol. Biol. 31:1–40.

Willhite, C.C. and Ferm, V.H. 1984. Prenatal and developmental toxicology of arsenicals. Adv. Exp. Med. Biol. 177:205–228.

Wilson, J.G. 1977. Current status of teratology: general principles and mechanisms derived from animal studies. In Handbook of Teratology, eds. J.G. Wilson and C.F. Fraser, p. 47. New York, NY: Plenum Press

Wlodarczyk, B.J., Bennett, G.D., Calvin, J.A. et al. 1996. Arsenic-induced neural tube defects in mice: alterations in cell cycle gene expression. Reprod. Toxicol. 10:447–454.

Yelin, R., Kot, H., Yelin, D. et al. 2007. Early molecular effects of ethanol during vertebrate embryogenesis. Differentiation 75:393–403.

Yon, J.M., Baek, I.J., Lee, S.R. et al. 2008. The spatio-temporal expression pattern of cytoplasmic Cu/Zn superoxide dismutase (SOD1) mRNA during mouse embryogenesis. J. Mol. Histol. 39:95–103.

Zaken, V., Kohen, R., Ornoy, A. 2000. The development of antioxidant defense mechanism in young rat embryos in vivo and in vitro. Early Pregnancy 4:110–123.

Zakeri, Z.F. and Ahuja, H.S. 1997. Cell death/apoptosis: normal, chemically induced, and teratogenic effect. Mut. Res. 396:149–161.

Chapter 4
Overview and Comparison of Idiopathic, Neuromuscular, and Congenital Forms of Scoliosis

Benjamin Alman

Scoliosis is really a physical finding, a lateral curvature of the spine. There are a number of potential causes, not all of which are related to a primary spinal deformity. For instance, if a patient has one leg longer than another, they will display a curved spine; otherwise their trunk would leave the pelvis at an angle, causing the individual to always look like they are leaning to one side. In a similar way, if an individual leans their back to one side, they will also show a scoliosis on a radiograph. Radiographs are often used to make the diagnosis, but because of these "postural" causes, scoliosis is usually defined as a curve on a standing radiograph of the spine measuring at least 10° (Binstadt et al. 1978, Cobb 1958).

The causes of scoliosis can be broadly classified as congenital, neuromuscular, syndrome related, idiopathic, and due to a secondary cause (Beals 1973). Another way to think about this is that scoliosis can be caused by a primary problem related to the spine itself (idiopathic and congenital scoliosis) or can be secondary to an underlying more systemic disorder (neuromuscular or syndromic scoliosis) (Table 4.1).

Congenital scoliosis is due to a vertebral malformation during fetal development, which results in a deviation of the normal spinal alignment. Neurologic conditions can cause curves due to muscle weakness or muscle imbalance, with muscles on one side of the spine pulling laterally more than muscles on the other side of the spine. This includes conditions such as cerebral palsy, paralysis, and Duchenne muscular dystrophy. These curves often have a typical radiographic appearance, with a long curve over most of the spine, often in a "c" shape, so that the patient is made off balance by the curve. Syndromes that are not associated with a neuromuscular problem, but instead are associated with a structural defect in the connective tissues can also cause scoliosis. This is presumably because gene mutation causes the bone and soft tissues to weaken, resulting in skeletal deformity. Several syndromes fall into this category including Marfan syndrome, osteogenesis imperfecta, and Ehlers–Danlos.

B. Alman (✉)
Hospital for Sick Children and University of Toronto, Toronto, ON, Canada
e-mail: benjamin.alman@sickkids.ca

K. Kusumi, S.L. Dunwoodie (eds.), *The Genetics and Development of Scoliosis*,
DOI 10.1007/978-1-4419-1406-4_4, © Springer Science+Business Media, LLC 2010

Table 4.1 Selected
syndromes and
neuromuscular conditions
associated with scoliosis

Achondroplasia
Arthrogryposis
Cerebral palsy
Charcot–Marie–Tooth disease
Congenital hypotonia
Duchenne muscular dystrophy
Ehlers–Danlos syndrome
Marfan syndrome
Myelomeningocele
Neurofibromatosis
Osteogenesis imperfecta
Paralysis
Poliomyelitis
Spinal muscular atrophy

In Ehlers–Danlos and Marfan, laxity of the soft tissue ligaments connecting the vertebra is associated with scoliosis. Osteogenesis imperfecta is associated with both soft tissue laxity and also weak bones that easily deform, and this condition can also cause scoliosis (Raff and Byers 1996).

The cause of idiopathic scoliosis, as the name suggests, is not known, although there is a wealth of data showing a familial occurrence (Kouwenhoven and Castelein 2008). Indeed, linkage studies suggest a genetic etiology, and several scoliosis loci have been identified. Idiopathic scoliosis behaves clinically in different ways depending on the age at which the curve presents. As such, it has been subdivided into adolescent, juvenile, and infantile forms. Many cases of infantile scoliosis and a few cases of juvenile scoliosis will resolve on their own. For the other idiopathic scoliosis patients, as a general rule, the younger a child presents, the more severe the ultimate degree of scoliosis that will develop.

Routine screening for scoliosis is no longer recommended in most jurisdictions (Weinstein et al. 2008). As such, patients usually present with a noticeable spinal deformity or, more likely, chest wall and back asymmetry. Whether identified by the patient, their parents, or through identification by a primary caregiver, posterior chest wall prominence is the most outward manifestation of spinal curvature. Other body characteristics may include shoulder asymmetry and overall posture imbalance in the coronal plane.

Sometimes pain can cause an individual to stand with their back bent and cause a scoliosis. This includes painful conditions such as infection, secondary to trauma, or malignancy. Brain tumors, intraspinal tumors, boney tumors, and extra-spinal malignancies, such as a retroperitoneal tumors, can all present with scoliosis. As such, the first priority for a physician caring for a patient with scoliosis is to evaluate for such serious, potential life threatening conditions (Janicki and Alman 2007).

The two reasons for a physician to become involved in the care of an individual with scoliosis are to identify the cause and to treat potentially deleterious effects of the curve. The medical evaluation of a patient with scoliosis includes a medical history, physical examination, and appropriate diagnostic radiographic tests. Much of the history and physical examination is focused on identifying more ominous causes of scoliosis. Young age at onset (younger than 10 years of age), rapid curve progression, and the presence of neurological symptoms are the most useful findings in identifying nonidiopathic causes of scoliosis.

Once scoliosis is suspected, spinal radiographs are usually obtained. These films are taken with a patient standing if possible. Curve magnitude is measured using a technique referred to as the Cobb method, in which the greatest angle between the end plates of the most tilted vertebra on each side of the apex of the curve is measured (Fig. 4.1). There is a high intra- and interobserver variability to this measurement. Studies show a 5° error of measurement in idiopathic cases, and studies report up to a 15° error of measurement in congenital cases. There are a number of clues that can be elucidated from the radiographs as to curve etiology. There should be two pedicles at every level, and absence of a pedicle suggests either a congenital, neoplastic, or infectious etiology. In idiopathic scoliosis, the apex of the curve usually points away from the heart, to the right, and there should be rotation of the spine, with the apex of the curve having the most rotation. Idiopathic scoliosis is not only a deformity in the coronal plane but also a rotational deformity. A scoliotic curve without rotation should be investigated for other causes. If back pain

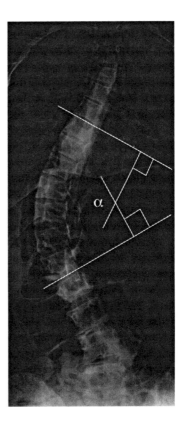

Fig. 4.1 Cobb angle (α) measurement on a standard anterior–posterior view spinal radiograph

is present, additional studies should be considered, such as a lateral radiograph of the spine including the lumbosacral region, to look for vertebral abnormalities associated with diagnoses such as spondylolysis (an idiopathic fracture of the posterior elements of the vertebral body), spondylolisthesis (a slipping forward of one vertebral body on the adjacent one), infection, or bony destruction. Further imaging for patients with scoliosis and back pain may include a bone scan or a magnetic resonance imaging (MRI) test. An MRI is not needed for the treatment of most patients with idiopathic scoliosis although some treating centers will request an MRI for any patient with scoliosis (Kotwicki 2008, Cassar-Pullicino and Eisenstein 2002, Carman et al. 1990, Morrissy et al. 1990).

In addition to the obvious differences in the etiology of the various types of scoliosis, each has differences in their clinical behavior and in the implications of spinal deformity (Howard et al. 2007, Mercado et al. 2007, Donaldson et al. 2007, Smith et al. 2006). Because radiographs are easy to observe, much of the literature on scoliosis outcome is based on the degree of curve progression. Despite this, there is a poor relationship between curve magnitude and the effect of scoliosis on an individual. Furthermore, depending on the location of a curve and an association with an underlying condition, curves of equal magnitude can have very different effects.

Several population studies have examined the natural history of curve progression in idiopathic, congenital, and neuromuscular scoliosis. These studies give an overall risk of progression in a population, but unfortunately, there is strong variability in how curves behave in individual patients, so that observation over time is still required to make treatment decision based on progression in individual patients. Further confounding issues of curve progression is the relatively poor reliability of measures of degree of scoliosis on radiographs. This means that one needs to measure at least a 5° change in the magnitude of the degree of curve to be considered a change (Carman et al. 1990, Morrissy et al. 1990).

Long-term studies of adolescent idiopathic scoliosis show little deleterious effect in most untreated patients. In patients with idiopathic scoliosis in whom surgery was recommended, but surgery was declined, there were little long-term differences in pain, physical function, general health, or occupation when compared to the general population. This makes it important to clearly define treatment goals in scoliosis. Indeed, these data provide the rationale for an ongoing prospective randomized trial of brace treatment versus no brace treatment for idiopathic scoliosis.

Severe curves centered in the thoracic spine can be associated with pulmonary function insufficiency, and changes in pulmonary function test results can be seen once a curve in the thoracic region exceeds 70°, and symptomatic changes are rare in curves measuring less than 90° (Barrios et al. 2005, Smyth et al. 1984, Weber et al. 1975). Some forms of congenital scoliosis are associated with more severe pulmonary compromise, and this is due to the overall short trunk height or to a severe curvature occurring in early infancy. Such deformities early in childhood can impede normal lung development, as there is insufficient space available for normal pulmonary development. In these cases there is often multiple rib abnormalities present which also contribute to the pulmonary insufficiency (Ramirez et al. 2007, Campbell et al. 2003). Patients with a neuromuscular etiology to their scoliosis often have poor pulmonary function. It is, however, not clear if the poor pulmonary function is related to the spinal deformity or to the underlying weakness of the muscles controlling respiration (Finder et al. 2004). As such, pulmonary effects from scoliosis are primarily limited to large curves, which occur most commonly in either idiopathic scoliosis that present at a very early age or severe cases of congenital scoliosis.

Pain can occur in all forms of scoliosis, but it is not clear that the incidence of pain is greater than in the general population. While there are no population studies of pain in congenital scoliosis, this type of spinal deformity does not seem to be associated with a higher frequency of pain than idiopathic scoliosis. In some cases of neuromuscular scoliosis or syndromic scoliosis pain may be associated with spinal deformity, particularly because sitting and positioning may be difficult and associated with abnormal pressure on the skin in patients with poor muscle control. Indeed, this symptom is a major reason that children with neuromuscular scoliosis undergo surgery (Mercado et al. 2007).

One consequence of spinal deformity is cosmesis. In moderate or severe scoliosis, one can see that the effected individual has a spinal deformity. Interestingly, the degree of spinal deformity on radiographs does not correlate with cosmetic ratings (Donaldson et al. 2007, Smith et al. 2006). For many individuals with idiopathic or congenital scoliosis, cosmetic concerns are a major reason they seek intervention. For children with congenital scoliosis, involvement of fewer vertebra often has worse cosmetic effects than involvement of multiple vertebrae. The site of the congenital deformity is also important, as hemivertebrae near the top and bottom of the spine have more dramatics cosmetic effects than congenital deformities in the middle of the spine (Goldberg et al. 2002a, b).

Congenital scoliosis is due to skeletal abnormalities of the spine that are present at birth. These anomalies, which can include multiple levels, are the result of and broadly classified as a failure of formation or a failure of segmentation (or both) during vertebral development. Because these spinal deformities are present in utero, they may be first identified on fetal ultrasound. In addition, the underlying cause can be associated with abnormalities of other organ systems in up to 40% of cases (Arlet et al. 2003). Thus, it is important to identify associated anomalies with a thorough evaluation of the neurological, cardiovascular, and genitourinary systems (Arlet et al. 2003). In contrast, idiopathic scoliosis is not associated with a higher incidence of other organ malformations than in the normal population. Syndromic scoliosis can be associated with other anomalies depending on the cause. For instance, Marfan syndrome is associated with cardiovascular anomalies.

Treatment is based on the type, age of the patient, progression of the curve, the long-term implications of the spinal deformity, the location of the curve, and type of anomaly. Treatment can be divided into the broad categories of observation, bracing, and surgery. Observation is used for small curves that are unlikely to progress to the point of causing symptomatic problems. Bracing has been shown to be effective in slowing scoliosis severity progression in mild to moderate-sized idiopathic curves in skeletally immature children. It has not been demonstrated to alter curve progression in other forms of scoliosis, or in curves in skeletally mature individuals. Despite this, bracing is sometimes used to hold the back in an overall straighter alignment. Surgery is used for more severe curves, and in most cases is spinal fusion using instrumentation to hold the back into a straighter alignment. While texts often list a degree of curvature as an indication for surgery, there are only relative, as opposed to absolute, indications for surgery. Natural history studies of patients with idiopathic scoliosis after skeletal maturity, found that curves less than 30° do not progress, while most curves of greater than 50° continue to progress. The progression is on average approximately 1° per year. These data are often used to suggest that curves greater than 50° require surgery as they will progress later in life. Not all curves actually progress (indeed almost a third of curves over 50° will not substantially increase in severity), and severe physiologic abnormalities have not been found in patients who had refused surgery with idiopathic scoliosis. In contrast severe curves, which are usually secondary to juvenile idiopathic scoliosis or congenital scoliosis can progress to a severity having deleterious physiologic effects and as such benefit from surgical intervention. Treatment for very young children with severe curves is rather problematic. In young children who still have substantial growth left, and for whom bracing is not an option, spinal fusion will shorten the length of the spine, and in the thoracic region this may decrease the space available for the lungs, resulting in impaired pulmonary function. In these individuals, non-fusion surgery is unusually advocated (Janicki and Alman 2007, Kim et al. 2009, Heary et al. 2008). This surgery involves either instrumentation for the spine without fusion or the use of instrumentation to distract the ribs. Non-fusion surgery techniques have less than ideal outcomes, and as such this is an area of intense current research into spine deformity management.

Scoliosis is a physical finding that can have a variety of etiologies. While there are some similarities between scoliosis of different etiologies, there are also substantial differences. As such, one needs to consider the etiology and natural history of the various forms of scoliosis to develop effective therapeutic approaches for patients with scoliosis.

References

Arlet, V., Odent, T., and Aebi, M. 2003. Congenital scoliosis. Eur. Spine J. 12(5):456–463.

Barrios, C. et al. 2005. Significant ventilatory functional restriction in adolescents with mild or moderate scoliosis during maximal exercise tolerance test. Spine 30(14):1610–1615.

Beals, R.K. 1973. Nosologic and genetic aspects of scoliosis. Clin. Orthop. Relat. Res. 93:23–32.

Binstadt, D.H., Lonstein, J.E., and Winter, R.B. 1978. Radiographic evaluation of the scoliotic patient. Minn. Med. 61(8):474–478.

Campbell, R.M., Jr. et al. 2003. The characteristics of thoracic insufficiency syndrome associated with fused ribs and congenital scoliosis. J. Bone Joint Surg. Am. 85-A(3):399–408.

Carman, D.L., Browne, R.H., and Birch, J.G. 1990. Measurement of scoliosis and kyphosis radiographs. Intraobserver and interobserver variation. J. Bone Joint Surg. Am. 72(3):328–333.

Cassar-Pullicino, V.N. and Eisenstein, S.M. 2002. Imaging in scoliosis: what, why and how? Clin. Radiol. 57(7):543–562.

Cobb, J.R. 1958. Scoliosis; quo vadis. J. Bone Joint Surg. Am. 40-A(3):507–510.

Donaldson, S. et al. 2007. Surgical decision making in adolescent idiopathic scoliosis. Spine 32(14):1526–1532.

Finder, J.D. et al. 2004. Respiratory care of the patient with Duchenne muscular dystrophy: ATS consensus statement. Am. J. Respir. Crit. Care Med. 170(4):456–465.

Goldberg, C.J. et al. 2002a. Growth patterns in children with congenital vertebral anomaly. Spine 27(11):1191–1201.

Goldberg, C.J. et al. 2002b. Long-term results from in situ fusion for congenital vertebral deformity. Spine 27(6):619–628.

Heary, R.F., Bono, C.M, and Kumar, S. 2008. Bracing for scoliosis. Neurosurgery 63(3 Suppl):125–130.

Howard, A. et al. 2007. Improvement in quality of life following surgery for adolescent idiopathic scoliosis. Spine 32(24):2715–2718.

Janicki, J.A. and Alman, B. 2007. Scoliosis: review of diagnosis and treatment. Paediatr. Child Health 12(9):771–776.

Kim, H.J., Blanco, J.S., and Widmann, R.F. 2009. Update on the management of idiopathic scoliosis. Curr. Opin. Pediatr. 21(1):55–64.

Kotwicki, T. 2008. Evaluation of scoliosis today: examination, X-rays and beyond. Disabil Rehabil. 30(10):742–751.

Kouwenhoven, J.W. and Castelein, R.M. 2008. The pathogenesis of adolescent idiopathic scoliosis: review of the literature. Spine 33(26):2898–2908.

Mercado, E., Alman, B., and Wright, J.G. 2007. Does spinal fusion influence quality of life in neuromuscular scoliosis? Spine 32(19 Suppl):S120–S125.

Morrissy, R.T. et al. 1990. Measurement of the Cobb angle on radiographs of patients who have scoliosis. Evaluation of intrinsic error. J Bone Joint Surg. Am. 72(3):320–327.

Raff, M.L. and Byers, P.H. 1996. Joint hypermobility syndromes. Curr. Opin. Rheumatol. 8(5):459–466.

Ramirez, N. et al. 2007. Natural history of thoracic insufficiency syndrome: a spondylothoracic dysplasia perspective. J. Bone Joint Surg. Am. 89(12):2663–2675.

Smith, P.L. et al. 2006. Parents' and patients' perceptions of postoperative appearance in adolescent idiopathic scoliosis. Spine 31(20):2367–2374.

Smyth, R.J. et al. 1984. Pulmonary function in adolescents with mild idiopathic scoliosis. Thorax 39(12):901–904.

Weinstein, S.L. et al. 2008. Adolescent idiopathic scoliosis. Lancet 371(9623):1527–1537.

Weber, B. et al. 1975. Pulmonary function in asymptomatic adolescents with idiopathic scoliosis. Am. Rev. Respir. Dis. 111(4):389–397.

Chapter 5
Abnormal Vertebral Segmentation (*or* Segmentation Defects of the Vertebrae) and the Spondylocostal Dysostoses

Peter D. Turnpenny

Introduction

In the most commonly occurring form of scoliosis in clinical practice, adolescent idiopathic scoliosis, there is no apparent malformation of the vertebrae. The process of *segmentation* that gives rise to somites, followed by *resegmentation* giving rise to vertebrae, and the subsequent *formation* of the vertebrae all appear normal. Abnormal formation of the vertebrae, for example, in some of the skeletal dysplasias that give rise to platyspondyly, may not necessarily be accompanied by scoliosis. In conditions with segmentation defects of the vertebrae (SDV), scoliosis is a frequent feature, though to a very variable and unpredictable degree. These conditions frequently present as *congenital* scoliosis. In clinical practice SDV are of interest to a variety of disciplines. Radiologists seek to describe abnormal imaging patterns, spinal surgeons make difficult decisions about surgery on affected children and adults, pediatricians care for the wider consequences such as respiratory insufficiency, and for clinical geneticists SDV are an important "handle" in the search for specific diagnoses, in making decisions concerning genetic testing, and in providing genetic risk counseling. SDV can occur as part of any one of an extensive range of syndromes (Table 5.1) as well as an isolated anomaly. This multiplicity of syndromes, with their different genetic and nongenetic causes, highlights the sensitivity and susceptibility of axial skeletal development to perturbations of normal somitogenesis. Apart from the Mendelian forms of SCD due to mutated genes in the Notch signaling pathway (see below), the function of other genes implicated in syndromes that include SDV is poorly understood. This includes, for example, *ROR2* in Robinow syndrome and *CHD7* in CHARGE syndrome. As a birth defect SDV

P.D. Turnpenny (✉)
Clinical Genetics Department, Royal Devon & Exeter Hospital and Peninsula Medical School, Exeter EX1 2ED, UK
e-mail: peter.turnpenny@rdeft.nhs.uk

K. Kusumi, S.L. Dunwoodie (eds.), *The Genetics and Development of Scoliosis*,
DOI 10.1007/978-1-4419-1406-4_5, © Springer Science+Business Media, LLC 2010

Table 5.1 Some syndromes that include segmentation defects of the vertebrae

Syndrome	OMIM reference	Gene(s)
Acrofacial dysostosis*	263750	
Aicardi*	304050	
Alagille	118450	*JAGGED1, NOTCH2*
Anhalt*	601344	
Atelosteogenesis III	108721	*FLNB*
Campomelic dysplasia	211970	*SOX9*
Casamassima–Morton–Nance*	271520	
Caudal regression*	182940	
Cerebro-facio-thoracic dysplasia*	213980	
CHARGE	214800	*CHD7*
"Chromosomal"		
Currarino	176450	*HLXB9*
De La Chapelle*	256050	
DiGeorge/Sedláčková	188400	Haploinsufficiency *22q11.2*
Dysspondylochondromatosis*		
Femoral hypoplasia-unusual facies*	134780	
Fibrodysplasia ossificans progressiva	135100	*ACVR1*
Fryns–Moerman*		
Goldenhar* (oculo-auriculo-vertebral spectrum)	164210	
Holmes–Schimke*		
Incontinentia pigmenti	308310	*NEMO*
Kabuki*	147920	
Kaufman–McKusick	236700	*MKKS*
KBG syndrome*	148050	
Klippel–Feil*	148900	? *PAX1*
Larsen	150250	*FLNB*
Lower mesodermal agenesis*		
Maternal diabetes*		
MURCS association*	601076	
Multiple pterygium syndrome	265000	*CHRNG*
OEIS syndrome*	258040	
Phaver*	261575	
Rapadilino	266280	*RECQL4*
Robinow	180700	*ROR2*
Rolland–Desbuquois*	224400	
Rokitansky sequence*	277000	? *WNT4*
Silverman	224410	*HSPG2*
Simpson–Golabi–Behmel	312870	*GPC3*
Sirenomelia*	182940	
Spondylocarpotarsal synostosis	269550	*FLNB*
Spondylocostal dysostosis	277300	*DLL3, MESP2, LFNG, HES7*
Spondylothoracic dysostosis*	277300	*MESP2*
Thakker–Donnai*	227255	

Table 5.1   (continued)

Syndrome	OMIM reference	Gene(s)
Toriello*		
Urioste*		
VATER/VACTERL*	192350	
Verloove-Vanhorick*	215850	
Wildervanck*	314600	
Zimmer*	301090	

*Underlying cause not known

may occur as frequently as 0.5–1.0/1000 births, though true incidence and prevalence figures are unknown (Wynne-Davies 1975, Purkiss et al. 2002). Many of the radiological phenotypes encountered do not match to a recognizable syndrome and the cause is frequently unidentified. Therefore, the clinical and research communities face enormous challenges in stratifying this broad group of conditions more accurately, understanding the natural history of the various disorders, and identifying the causes. This chapter will concentrate on the small but well-defined group of Mendelian conditions to which the term spondylocostal dysostosis (SCD) has been applied.

The Problem of Nomenclature

In describing segmentation abnormalities of the spine and ribs, the terms *Jarcho–Levin syndrome, costovertebral/spondylocostal/spondylothoracic dysostosis/dysplasia* all feature in the medical literature. Strictly speaking these disorders are *dysostoses* rather than *dysplasias*. A *dysplasia* refers to a developmental and ongoing abnormality of chondro-osseous tissues during prenatal and postnatal life, while a *dysostosis* is a stable condition resulting from a formation abnormality early in morphogenesis. In SDV, radiological features are crucial in syndrome delineation, which in turn is essential for offering accurate genetic counseling. However, the use of nomenclature in the medical literature is inconsistent and confusing, and this problem is widely, but understandably, reflected in clinical practice. Multiple terms are applied to the same phenotype, and a great diversity of phenotypes may come under the same term, particularly *Jarcho–Levin syndrome* (JLS). Some authors have recognized the existence of different entities and applied a rational distinction in the use of terms (Aymé and Preus 1986, Roberts et al. 1988) but there is a need for a systematic approach, which will be addressed in more detail later.

The eponymous *Jarcho–Levin syndrome* (JLS) is frequently used as an umbrella term across the entire spectrum of radiological phenotypes that include SDV and abnormal rib alignment. In 1938, Jarcho and Levin reported two siblings of Puerto ican Rican origin with SDV of the entire vertebral column, though most severe in the thoracic region (Jarcho and Levin 1938). Fusion of several ribs was present and both subjects died in infancy of respiratory failure. Close scrutiny suggests the phenotype is closest to the form of SCD that we now call "type 2," due to mutations in *MESP2* (see below). To prevent confusion we believe it is preferable to avoid the use of the eponymous JLS. *Klippel–Feil anomaly* (KFA), another frequently used eponym in relation to cervical vertebral fusion anomalies, is more specific and therefore more useful, though even here a great diversity of phenotypes occur with fusion of the cervical vertebrae. A rarer eponymous entity with marked diversity in the few reported cases is the *Casamassima–Morton–Nance syndrome* (Casamassima et al. 1981), which combines SDV with urogenital anomalies, apparently following autosomal recessive inheritance. However, subsequent reports (Daikha-Dahmane et al. 1998, Poor et al. 1983) demonstrated a different SDV phenotype from the cases of Casamassima et al. and consistency across all three reports, based on the SDV phenotype, is lacking.

In the early literature, the term *costovertebral dysplasia* can be found (Norum and McKusick 1969, Cantú et al. 1971, David and Glass 1983) but is less often used today. There is a preference for *spondylocostal dysostosis* (SCD) for those phenotypes with extensive segmental vertebral involvement (≥ 10 contiguous segments plus rib abnormalities with points of fusion), and *multiple* SDV when >1 and <10 vertebral segments are involved (Bartsocas et al. 1981) (Table 5.2). However, in practice a consensus on the use of terminology has yet to be achieved. As with JLS, the term SCD is used for a wide variety of abnormal axial radiological phenotypes, including those with gross asymmetry that are mostly sporadic (Fig. 5.1).

The term *spondylothoracic dysostosis/dysplasia* (STD) was first proposed by Moseley and Bonforte (1969) and is best reserved for the distinctive condition, most commonly reported in Puerto Ricans, characterized by severe trunkal shortening

Table 5.2 Proposed definitions for the terms spondylocostal dysostosis (SCD) and spondylothoracic dysostosis (STD)

Features	Spondylocostal dysostosis (SCD)	Spondylothoracic dysostosis (STD)
General	No major asymmetry to chest shape Mild, nonprogressive scoliosis Multiple SDV (MSDV) ≥ 10 contiguous segments Absence of a bar Mal-aligned ribs with intercostal points of fusion	Chest shape symmetrical, with ribs fanning out in a "crab-like" appearance Mild, nonprogressive scoliosis, or no scoliosis Generalized SDV (GSDV) Regularly aligned ribs, fused posteriorly at the costovertebral origins, but no points of intercostal fusion
Specific, descriptive	"Pebble beach" appearance of vertebrae in early childhood radiographs (Fig. 5.3)	"Tramline" appearance of prominent vertebral pedicles in early childhood radiographs, not seen in SCD "Sickle cell" appearance of vertebrae on transverse imaging (Cornier et al. 2004)

Fig. 5.1 An SCD phenotype
demonstrating gross
asymmetry of the thoracic
cage. Cases like this do not
appear to follow Mendelian
inheritance. From Turnpenny
et al. 2007

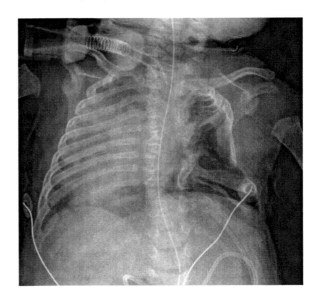

Fig. 5.1 An SCD phenotype demonstrating gross asymmetry of the thoracic cage. Cases like this do not appear to follow Mendelian inheritance. From Turnpenny et al. 2007

and a radiological appearance of the ribs fanning out from a crowded vertebro-costal origin in a "crab-like" fashion. The ribs appear fused posteriorly, though otherwise they are neatly aligned and packed tightly together, and in contrast to the appearance in SCD, they show no points of fusion other than that at their posterior origins. The thoracic vertebrae appear most severely affected and "telescoped" together, and scoliotic curves in early life are uncommon. This distinctive phenotype has been well characterized in recent studies (Cornier et al. 2000, Cornier et al. 2004). Infant mortality is approximately 50% due to restrictive respiratory insufficiency, though survival may depend on the quality of medical care available. As illustration of the confusion in nomenclature, a number of case reports in the literature that demonstrate the typical STD phenotype, predominantly in Puerto Ricans, have been reported as examples of JLS (Moseley and Bonforte 1969, Lavy et al. 1966, Pochaczevsky et al. 1971, Pérez-Comas and García-Castro 1974, Gellis and Feingold 1976, Trindade and de Nóbrega 1977, Solomon et al. 1978, Tolmie et al. 1987, Schulman et al. 1993, McCall et al. 1994, Mortier et al. 1996). Other case reports (Karnes et al. 1991, Simpson et al. 1995, Aurora et al. 1996, Eliyahu et al. 1997, Rastogi et al. 2002) designated as JLS are neither very similar to those described by Jarcho and Levin (Jarcho and Levin 1938) nor consistent with STD.

A number of attempts have been made to define the use of existing nomenclature and arrive at some form of classification. The scheme proposed by Mortier et al. (1996) combines phenotype and inheritance pattern (Table 5.3). The scheme proposed recently by Takikawa et al. (2006) allows a very broad definition of SCD (Table 5.4), but both these schemes identify JLS with a "crab-like" chest. McMaster's surgical approach to classification distinguishes between formation and segmentation errors (Table 5.5, McMaster and Singh 1999). KFA, used to describe

Table 5.3 Previously proposed classifications for SDV classification of SDV according to Mortier et al. (1996)

Nomenclature	Definition
Jarcho–Levin syndrome	Autosomal recessive
	Symmetrical crab-like chest, lethal
Spondylothoracic dysostosis	Autosomal recessive
	Intrafamilial variability, severe/lethal
	Associated anomalies uncommon
Spondylocostal dysostosis	Autosomal dominant
	Benign
Heterogeneous group	Sporadic
	Associated anomalies common

Table 5.4 Previously proposed classification/definition of SDV according to Takikawa et al. (2006)

Nomenclature	Definition
Jarcho–Levin syndrome	Symmetrical crab-like chest
Spondylocostal dysostosis	≥ 2 vertebral anomalies associated with rib anomalies (fusion and/or absence)

Table 5.5 Previously proposed classifications for SDV: Classification (surgical/anatomical) of vertebral segmentation abnormalities causing congenital kyphosis/kyphoscoliosis, according to McMaster and Singh (1999)

Type	Anatomical deformity	Anomalies
I	Anterior failure of vertebral body formation	Posterolateral quadrant vertebrae
		– single vertebra
		– two adjacent vertebrae
		Posterior hemivertebrae
		– single vertebra
		– two adjacent vertebrae
		Butterfly (sagittal cleft) vertebrae
		Anterior or anterolateral wedged vertebrae
		– single vertebra
		– two adjacent vertebrae
II	Anterior failure of vertebral body segmentation	Anterior unsegmented bar
		Anterolateral unsegmented bar
III	Mixed	Anterolateral unsegmented bar contralateral posterolateral quadrant vertebrae
IV	Unclassifiable	

different forms of cervical vertebral fusion or segmentation error, has been subclassified (Table 5.6, Feil 1919, Thomsen et al. 1997), and Clarke et al. (1998) proposed a further, detailed classification combining modes of inheritance (Table 5.7).

Table 5.6 Classification of Klippel–Feil anomaly, referring to segmentation defects or fusion of the cervical vertebrae, according to Feil (1919) and Thomsen et al. (1997)

Type	Site	Anomaly
I	Cervical and upper thoracic	Massive fusion with synostosis
II	Cervical	One or two interspaces only, hemivertebrae, occipito-atlantoid fusion
III	Cervical and lower thoracic or lumbar	Fusion

Table 5.7 Classification of Klippel–Feil anomaly, adapted from Clarke et al. (1998)

Class	Vertebral fusions	Inheritance	Possible anomalies	Overlap with other classifications
KF1	Only class with C1 fusion Variable expression of other fusions	Recessive	Very short neck, heart, urogenital, craniofacial, hearing, likmb, digital, oclar defects; variable	Types I, II, and III (Klippel and Feil)
KF2	C2-3 dominant C2-3 most rostral fusion Cervical, thoracic, and lumbar fusion variable within a family	Dominant	Craniofacial, hearing, otolaryngeal, skeletal, and limb defects; variable	Types I, II, and III (Klippel and Feil)
KF3	Isolated cervical fusions Any cervical fusion except C1-2	Recessive or reduced penetrance	Craniofacial dysmorphology; variable	Type II (Klippel and Feil)
KF4	Fusion of cervical vertebrae, limited data	Possibly X-linked; predominantly females	Hearing and ocular anomalies; possibly heart defects	Commonly termed Wildervanck syndrome

In reality, the use of a limited number of terms fails to reflect the great diversity of radiological SDV phenotypes seen in clinical practice. Furthermore, segmentation disorders and their diversity are not fully catered for within the classification of osteochondrodysplasias (Offiah and Hall 2003, Superti-Furga and Unger 2007).

The Diversity of Phenotypes Manifesting Axial Skeletal Defects

In clinical practice sporadically occurring cases of SDV are far more common than familial, and sporadic cases are more likely to be associated with additional anomalies. There is wide clinical heterogeneity within the sporadic group (Martínez-Frías et al. 1994) and the literature was well reviewed by Mortier et al. (1996). Anal and urogenital anomalies occur most frequently (Poor et al. 1983, Pochaczevsky et al. 1971, Tolmie et al. 1987, McCall et al. 1994, Karnes et al. 1991, Bonaime et al. 1978, Eller and Morton 1976, Devos et al. 1978, Kozlowski 1984, Giacoia and Say 1991, Murr et al. 1992, Lin and Harster 1993) followed by a variety of congenital heart disease (Mortier et al. 1996, Aurora et al. 1996, Kozlowski 1984, Delgoffe et al. 1982, Ohzeki et al. 1990). Limb abnormalities occur but are generally of a minor nature, e.g., talipes and oligodactyly or polydactyly (Mortier et al. 1996, Karnes et al. 1991). Infrequently, diaphragmatic hernia is a feature (Martínez-Frías et al. 1994). As a minor anomaly, inguinal and abdominal herniae are frequently reported in association with SDV.

Many case reports could reasonably be assigned a diagnosis of the VATER (Vertebral defects, Anal atresia, Tracheo-Esophageal fistula, Radial defects, and Renal anomalies) or VACTERL (Vertebral defects, Anal atresia, Cardiac defects, Tracheo-Esophageal fistula, Radial defects and Renal anomalies, non-radial Limb defects) associations (Kozlowski 1984) and this would appear to be a very heterogeneous group with few clues regarding causation at the present time. A frequent association with severe SDV, affecting rib number and alignment, is neural tube defect (NTD, Wynne-Davies 1975, Martínez-Frías et al. 1994, Eller and Morton 1976, Kozlowski 1984, Giacoia and Say 1991, McLennan 1976, Naik et al. 1978, Lendon et al. 1981, Sharma and Phadke 1994). This NTD-associated group should be classified separately from the SCD group because the primary developmental pathology presumably lies in the processes determining neural tube closure as distinct from somitogenesis, and it is valid to classify on the basis of underlying cause rather than the radiological features (Martínez-Frías 1996). Similarly, an association between spina bifida occulta, and/or diastematomyelia, and SDV has been reported (Aymé and Preus 1986, Poor et al. 1983, Herold et al. 1988, Reyes et al. 1989), strongly suggesting a causal link or sequence, though the mechanisms remain to be elucidated.

SDV may be a consequence of maternal diabetes syndrome, which can give rise to multiple congenital anomalies. Classically, the malformation is caudal regression to a varying degree, i.e., absent sacrum or agenesis of the lower vertebral column (Bohring et al. 1999), features that overlap with the ill-defined disorder known as axial mesodermal dysplasia. However, there are patients with hemivertebrae (Novak and Robinson 1994) and various forms of axial skeletal defects (Fig. 5.2) following poorly controlled diabetes in pregnancy in an insulin-dependent mother.

Some clues to genetic causes of SDV may come from patients with axial skeletal defects and chromosomal abnormalities. These cases are relatively rare and apart from trisomy 8 mosaicism (Riccardi 1977) there is no clear consistency to the group with chromosome abnormalities. Deletions affecting both 18q (Dowton et al. 1997) and 18p (Nakano et al. 1977) have been reported, a supernumerary dicentric 15q marker (Crow et al. 1997) and an apparently balanced translocation between chromosomes 14 and 15 (De Grouchy et al. 1963). It can be postulated that haploinsufficiency may unmask a new Mendelian locus for SCD but the paucity and

Fig. 5.2 Abnormal segmentation in a child whose mother had poorly controlled diabetes mellitus in pregnancy. This is a recognizable pattern in maternal diabetes syndrome

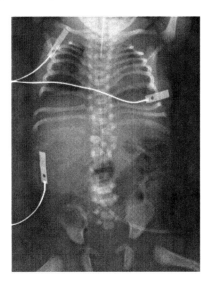

diversity of these cases may indicate that the resulting SDV in association with multiple congenital anomalies represents a common pathway of complex pleiotropic developmental mechanisms that are sensitive to a range of unbalanced karyotypes. It is also possible that chromosome mosaicism accounts for some cases where there is marked asymmetry in the radiological phenotype, which would also explain sporadic occurrence, but skin or tissue biopsy is rarely undertaken.

Notch Signaling Pathway Genes and Spondylocostal Dysostosis

Notch signaling is a key cascade pathway in somitogenesis, and the function of many genes and their products, together with their complex interactions, has been at least partially elucidated through research in animal models. The consequences of many different heterozygous and homozygous gene knock-outs in animal models have been well described elsewhere in this book. In man, the functions of orthologous genes and their proteins obviously cannot be studied in the same way. Nevertheless, the association of a number of genes in the pathway with certain phenotypes has become established through linkage studies carried out on affected families, followed by candidate gene sequencing to identify pathogenic mutations. The diseases that result from mutations in Notch signaling genes are diverse and would not readily be considered to link to a common pathway. The affected organ systems include the vascular and central nervous systems; the skeleton, face, and limb; hematopoiesis; the determination of laterality; and the liver, heart, kidney, and eye. The reader is directed to articles that review the many functions of Notch signaling during development (Joutel and Tournier-Lasserve 1998, Gridley 1997, Gridley 2003, Pourquié and Kusumi 2001, Harper et al. 2003). As this chapter deals primarily with SDV, consideration is now given to the Notch pathway genes *DLL3*, *MESP2*, *LFNG,* and *HES7* genes, all of which give rise to autosomal recessive traits in man.

Delta-Like 3 (DLL3) and SCD

The key breakthrough in the search for a gene linked to SCD came with autozygosity mapping studies in a large inbred Arab kindred with seven affected individuals (Turnpenny et al. 1991, Turnpenny et al. 1999). The locus identified, 19q13.1, is syntenic with mouse chromosome 7 (Giampietro et al. 1999), which harbors the *Dll3* gene. A mutation in this gene was shown to be the cause of SDV in a radiation-damaged mouse known as *pudgy* (Dunwoodie et al. 1997, Grüneberg 1961, Kusumi et al. 1998). *DLL3* was therefore the obvious candidate for the SCD families demonstrating linkage to 19q13.1 and sequencing initially identified mutations in three consanguineous affected families, including the original large Arab kindred (Bulman et al. 2000). Approximately 30 mutations have now been identified in affected individuals (Turnpenny et al. 2007 and unpublished data) and *DLL3*-associated SCD is referred to as SCD type 1. The majority of affected subjects have been the offspring of consanguineous partnerships with Middle Eastern or Pakistani origins.

Human *DLL3* encodes a ligand for Notch signaling which has recently been shown to act as a negative regulator of Notch signaling (Ladi et al. 2005, Geffers et al. 2007). The gene comprises eight exons and spans approximately 9.2 kb of chromosome 19. A 1.9 kb transcript encodes a protein of 618 amino acids. The protein consists of a signal sequence, a delta-serrate-lag2 (DSL, receptor interacting) domain, six epidermal growth factor (EGF)-like domains, and a transmembrane domain (TM). Pathological mutations occur throughout all domains of the gene, three-quarters of which are protein truncating, the remainder being missense. Some missense mutations appear to give rise to a slightly milder phenotype and protein modeling studies may help explain these differential effects in due course. Studies in animal models have found that *DLL3* shows spatially restricted patterns of expression during somite formation and it is believed to have a key role in the cell signaling processes giving rise to somite boundary formation, which proceeds in a rostral–caudal direction with a precise temporal periodicity driven by an internal oscillator, or molecular "segmentation clock" (McGrew and Pourquié 1998, Pourquié 1999).

Mutated *DLL3* results in SDV throughout the entire spine with *all* vertebrae losing their normal form and regular three-dimensional shape. The most dramatic changes, radiologically, affect the thoracic vertebrae, and the ribs are mal-aligned with a variable number of points of fusion along their length (Fig. 5.3a). There is an overall symmetry of the thoracic cage despite some cases showing minor scoliotic curves that are nonprogressive and therefore do not usually require surgery. In early childhood, before ossification is complete, the vertebrae have smooth, rounded outlines – especially in the thoracic region – and for this we have suggested the

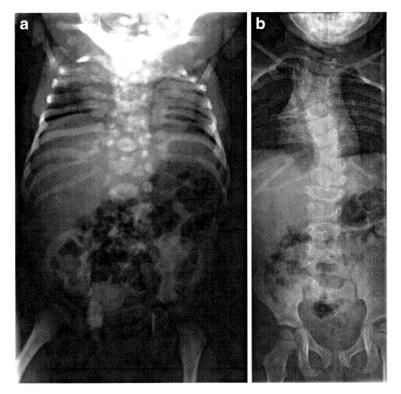

Fig. 5.3 a. The radiological phenotype of SCD type 1 due to mutated *DLL3*. For this pattern of rounded vertebrae with smooth edges the term "pebble beach sign" has been suggested. From Turnpenny et al. 2007. **b**. A milder phenotype due to a missense mutation in *DLL3*. Segmentation is less severe and there are mild scoliotic curves

term "pebble beach sign." Additional anomalies appear to be rare. However, in one case abdominal *situs inversus* was present but the link with mutated *DLL3* is uncertain. In another family, affected siblings homozygous for the exon 8 mutation 1369delCGCTCCCGGCTACATGG (C655M660del17) manifested a form of distal arthrogryposis in keeping with fetal akinesia sequence, and both succumbed in early childhood (C. McKeown, personal communication). Multiple inbreeding occurred in this particular family and it is possible that a separate autosomal recessive condition segregated coincidentally to SCD. Spinal cord compression and associated neurological features have not been observed, and intelligence and cognitive performance are normal. This suggests that *DLL3* is not required for brain development or function in humans, which contrasts to findings in the mouse, where central nervous system defects have been found (Pourquié 1999), including defects in the neuroventricles of the *pudgy* mouse (Kusumi et al. 2001, Dunwoodie et al. 2002).

Some human mutations have been identified in more than one family and haplotype studies have suggested common ancestry for at least two of these. These are (i) two ethnic Pakistani kindreds originating from Kashmir (949delAT mutation) and (ii) an ethnic Lebanese Arab family and ethnic Turkish family (614insGTC-CGGGACTGCG mutation). The 593insGCGGT (S198ins5) mutation is present in the original Arab kindred, homozygous in those affected, but in a Spanish family the affected child is heterozygous for the same mutation. Haplotype analysis on these two pedigrees does *not* support a common ancestry and the 593ins-GCGGT mutation is therefore believed to be recurrent, occurring as it does within a region of the gene with multiple repeat GCGGT sequences. Slipped mispairing during DNA replication is the likely explanation of this insertion mutation. Similarly, the missense mutation G504D has been found in northern European families (England and The Netherlands) and haplotype analysis suggests that it has occurred recurrently. One family with this mutation was originally reported as demonstrating autosomal dominant SCD (Floor et al. 1989). However, the family was shown to demonstrate pseudo-dominant inheritance (Whittock et al. 2004a). The mutations, C309R and G404C, are missense and appear to be associated with a slightly milder phenotype (Fig. 5.3b). G404C has been identified in two families from The Netherlands, in one of which the patients were homozygous, and in the other heterozygous, the second allele being G504D. Other missense mutations give rise to a phenotype indistinguishable from that caused by protein truncation mutations.

At present there is no confirmed case of AD SCD due to mutated *DLL3*. However, in the large Arab family reported by Turnpenny et al. (1991) one female heterozygous for the 593insGCGGT (S198ins5) mutation had a mild thoracic scoliosis but no associated segmentation abnormality in the thoracic region, and a very localized segmentation anomaly in the lower lumbar vertebrae. It is possible her scoliosis and lumbar segmentation anomaly were coincidental to her *DLL3* carrier status, as no other obligate carrier in the kindred is known to have had similar features.

In general, the consistency of the radiological phenotype in mutation positive cases (Turnpenny et al. 2003) means that, with experience, scrutiny of the radiograph usually makes it possible to identify those patients who will prove to have *DLL3* mutations. *DLL3* mutations have not been found in the wide variety of more common, though diverse, phenotypes that include SDV and abnormal ribs (Maisenbacher et al. 2005, Giampietro et al. 2006). Therefore, it appears that there is remarkable clinical homogeneity for the axial skeletal malformation due to mutated *DLL3*, which has significant implications for the application of genetic testing in the clinical setting.

In relation to defects of the axial skeleton in man, identification of the *DLL3* gene in SCD has represented a breakthrough in understanding the causative basis of this group of malformations, as well as highlighting another example of cross-species biological homology. It has become the paradigm for searching for the genetic basis of other SCD phenotypes, which led directly to the identification of *MESP2*, *LFNG*, and *HES7* in cases of SCD.

Mesoderm Posterior 2 (MESP2) and SCD

The identification of mutated *MESP2* in association with SCD arose from the study of a family of Lebanese Arab origin, with two affected children, in whom no *DLL3* mutations were found and linkage to 19q13.1 was excluded. In this family, the radiological phenotype was subtly different from SCD type 1. Generalized SDV was present, though with somewhat more angular features to the vertebrae compared to the appearance in *DLL3*-associated SCD (Fig. 5.4a). Genome-wide homozygosity mapping revealed six homozygous markers concentrated in a block on 15q (D15S153 to D15S120). Subsequent mapping further supported linkage to the 15q markers D15S153, D15S131, D15S205, and D15S127, and fine mapping demonstrated a 36.6 Mb region on 15q21.3–15q26.1, between markers D15S117 and D15S1004, with a maximum two-point lod score of 1.588 at $\theta = 0$. The region contains an excess of 50 genes and is syntenic to mouse chromosome 7, which includes the *Mesp2* gene. The *Mesp2* knockout mouse manifests altered rostro-caudal polarity, resulting in axial skeletal defects (Saga et al. 1997). The

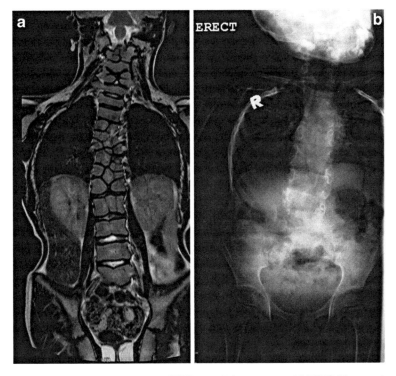

Fig. 5.4 a. The radiological phenotype of SCD type 2 due to mutated *MESP2*. The vertebrae are more angular than seen in SCD type 1 and the lumbar vertebrae are relatively spared. This case was homozygous for the 500–503dup mutation. From Whittock et al. 2004. **b.** Another SCD type 2 case. This patient was heterozygous for mutations K91E (c.271A > G) and I129F (c.385A > T)

predicted human gene, *MESP2*, comprises two exons spanning approximately 2 kb of genomic DNA at 15q26.1. Direct sequencing of *MESP2* gene in the two affected siblings demonstrated a homozygous 4-bp (ACCG) duplication mutation in exon 1, termed 500–503dup (Whittock et al. 2004b). The parents were heterozygotes and the unaffected sibling homozygous normal, consistent with the duplication segregating with SCD in the family. Fluorescent PCR excluded this mutation from 68 ethnically matched control chromosomes. The insertion is predicted to interrupt splicing, leading to a frameshift at the same point in the MESP2 protein. A further case with the same mutation and similar pattern of SDV has been identified (Bonafé and Superti-Furga, International Skeletal Dysplasia Society meeting 2005, Martigny) as well as a compound heterozygote case with novel mutations K91E (c.271A > G) and I129F (c.385A > T) (P Turnpenny, personal communication) (Fig. 5.4b).

The *MESP2* gene is predicted to produce a transcript of 1,191 bp encoding a protein of 397 amino acids with a predicted molecular weight of 41,744 Da and isoelectric point (pI) of 7.06. The human MESP2 protein has 58.1% identity with mouse MesP2, and 47.4% identity with human MESP1. Human *MESP2* amino terminus contains a basic helix–loop–helix (bHLH) region encompassing 51 amino acids divided into an 11-residue basic domain, a 13-residue helix I domain, an 11-residue loop domain, and a 16-residue helix II domain. The loop region is slightly longer than that found in homologues such as *paraxis*. The length of the loop region is conserved between *MESP1* and *MESP2* in mouse and human, and orthologues in *Xenopus* and chick. In addition, both *MESP1* and *MESP2* contain a unique CPXCP motif immediately carboxy-terminal to the bHLH domain. The amino- and carboxy-terminal domains are separated in human *MESP2* by a GQ repeat region also found in human *MESP1* (two repeats) but expanded in human *MESP2* (13 repeats). Mouse *MesP1* and *MesP2* do not contain GQ repeats but they do contain two QX repeats in the same region: mouse *MesP1* QSQS; mouse *MesP2* QAQM. Hydrophobicity plots indicate that *MesP1* and *MesP2* share a carboxy-terminal region that is predicted to adopt a similar fold, although *MesP2* sequences contain a unique region at the carboxy-terminus.

Sequence analysis of 20 ethnically matched and 10 non-matched individuals revealed the presence of a variable length polymorphism in the GQ region of human *MESP2*, beginning at nucleotide 535. This region contains a series of 12 bp repeat units. The smallest GQ region detected contains two type A units (GGG CAG GGG CAA, encoding the amino acids GQGQ), followed by two type B units (GGA CAG GGG CAA, encoding GQGQ), and one type C unit (GGG CAG GGG CGC, encoding GQGR). Analysis of this polymorphism in the matched and non-matched controls revealed allele frequencies that were not significantly different statistically between the two groups.

Thus it was shown that the frameshift mutation in the human homologue, *MESP2*, results in autosomal recessive SCD in man, and this was designated SCD type 2, or *MESP2*-associated SCD. *MESP2* is a member of the bHLH family of transcriptional regulatory proteins essential to a vast array of developmental processes (Massari and Murre 2000).

In murine somitogenesis *Mesp2* has a key role in establishing rostro-caudal polarity by participating in distinct Notch signaling pathways (Takahashi et al. 2000, Takahashi et al. 2003). *Mesp2* expression is induced by *Dll1*-mediated Notch signaling (presenilin1-independent) and *Dll3*-mediated Notch signaling (presenilin1-dependent), while inhibition of *Mesp2* expression is achieved through presenilin1-independent *Dll3*-Notch signaling. Since *Mesp2* can inhibit *Dll1* expression, this complex signaling network results in stripes of *Dll1*, *Dll3*, and *Mesp2* gene expression in the anterior PSM.

The extent to which the murine model directly correlates with somitogenesis in man is unknown, except that the phenotypes of the *Mesp2* and *Dll3* mutant mice closely resemble human SCD (Kusumi et al. 1998, Dunwoodie et al. 2002, Saga et al. 1997). Mutated *MESP2* that gives rise to the similar but distinctive phenotype, *spondylothoracic dysostosis* (STD), is dealt with later.

Lunatic Fringe (LFNG) and SCD

The identification of Notch pathway genes as causes of human SCD inevitably led to the hypothesis that other genes of the pathway may be implicated in other forms of generalized SDV. Among these the lunatic fringe (*LFNG*) gene, located on chromosome 7p22, was considered a strong candidate because the *Lfng*-null mouse has a nonlethal phenotype that includes costo-vertebral anomalies. The expression pattern of *Lfng* is oscillatory within the presomitic mesoderm and it is one of several "cycling" genes that are crucial to the integrity of the segmentation clock. *LFNG* encodes a glycosyltransferase that posttranslationally modifies the Notch family of cell surface receptors, a key step in the regulation of this signaling pathway (Haines and Irvine 2003) The LFNG protein is a fucose-specific β 1,3 *N*-acetylglucosaminyltransferase (Haines and Irvine 2003, Moloney et al. 2000) that functions in the Golgi to posttranslationally modify the Notch receptors, altering their signaling properties (Bruckner et al. 2000, Haines and Irvine 2003). Earlier studies have shown that *Lfng* gene expression is severely disregulated in *Dll3*-null mice, suggesting that *Lfng* expression is dependent on Dll3 function (Dunwoodie et al. 2002, Kusumi et al. 2004).

Initially, mutation screening in a series of affected subjects with diverse radiological phenotypes failed to identify any positive cases. Eventually another case was studied – an adolescent boy originating from northern Lebanon, the second of five children born to consanguineous parents. He had a short neck and trunk at birth and at 15 years had marked shortening of the thorax with a pectus carinatum and kyphoscoliosis. A spinal MRI confirmed multiple SDV in the cervical and the thoracic spine with a serpentine curve in the cervical spine, concavity to the right in the upper spine, and concavity to the left inferiorly. The thoraco-lumbar spine showed multiple hemivertebrae (Fig. 5.5) and there were also multiple rib anomalies. The spinal cord was normal with no evidence of a syrinx. At 15 he had a markedly short trunk with a height of 155 cm (5th percentile), lower segment of 92.5 cm, and an

Fig. 5.5 The radiological phenotype of SCD type 3 due to mutated *LFNG*. Only one case has yet been identified. The spinal shortening was very marked and all regions affected. From Sparrow et al. 2006

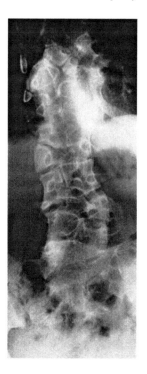

arm-span of 186.5 cm. As span:height ratio is close to unity normally, his stature was reduced by approximately 30 cm. At birth he had been noted to have a contracture of the left index finger and at age 15 he had hypoplasia of all the distal interphalangeal joints of the fingers, which were long and slender. Radiographs of the wrists and ankles were normal and the link between the spinal malformation and mild digital contractures remains speculative.

On sequencing the entire coding region and splice sites of the *LFNG* gene, a homozygous missense mutation (c.564C > A) in exon 3 was detected (Sparrow et al. 2006), resulting in substitution of leucine for phenylalanine (F188L). The proband's parents, with normal spinal anatomy, were heterozygous for the mutant allele. The phenylalanine residue substituted is highly conserved (Correia et al. 2003) and close to the active site of the enzyme. The mutation created a novel MseI restriction enzyme site, which was used to confirm the sequencing results in the pedigree. This variant was not found in 34 ethnically matched control subjects (68 chromosomes), and the underlying base substitution was not present in the NCBI SNP database (www.ncbi.nlm.nih.gov). Examination of the mutation within a fringe model based on solved glycosyltransferase structures showed the conserved phenylalanine residue (F188) to be located in a helix that packs against the strand containing Mn^{2+}-ligating residues, rather than being directly involved in UDP-*N*-acetylglucosamine or protein binding.

Further evidence of causality was provided by functional assays of the *LFNG* mutant. Two F187L mutations in mouse *Lfng* that correspond to F188L in human *LFNG* were generated: a c.564C > A mutation that encodes the rare leucine codon (TTA) observed in the proband and the [c.562T > C + c.567C > G] mutation encoding the most common human leucine codon (CTG). In addition, a previously characterized enzymatically inactive form of Lfng (D202A) was created that disrupts the conserved DDD Mn^{2+}-binding active site (Chen et al. 2001). Protein expression studies showed that both F187L mutant Lfng proteins were expressed at higher levels than wild-type or D202A Lfng forms, indicating that both translation efficiency and protein stability were not adversely affected by the F187L amino acid change. As Lfng protein is normally present in the Golgi apparatus, intracellular protein localization was examined using immunofluoresence. Wild-type and D202A mutant Lfng were localized predominantly to the Golgi while the F187L mutant Lfng did not co-localize with a marker. It was concluded that the F187L mutant form of Lfng was expressed but abnormally localized within the cell.

This single case of SCD due to mutated *LFNG* has provided further evidence that proper regulation of the Notch signaling pathway is an absolute requirement for correct patterning of the axial skeleton, at the same time defining SCD type 3, or *LFNG*-associated SCD.

Hairy and Enhancer of Split (HES7) and SCD

As with *LFNG*, the *HES7* gene was a candidate for being associated with an SCD phenotype because of its known role as a cycling gene in somitogenesis in experimental models such as mouse, chick, zebrafish, and *Xenopus*. In *Hes7*-null mice the somites do not segment normally and display disruption of anterior–posterior polarity (Bessho et al. 2001). Hes7, like Mesp2, is a protein of the basic helix–loop–helix (bHLH) superfamily of more than 125 DNA-binding transcription factors that are involved in a vast array of developmental processes. When compared to the others, HES (and HEY) proteins have an additional characteristic conserved domain that is immediately carboxy-terminal to the bHLH domain, referred to as the *Orange* domain, as well as a carboxy-terminal tetrapeptide. The oscillatory expression pattern of *Hes* genes in mice is achieved by an autoregulatory feedback loop. Following translation, HES proteins bind with their own promoters to repress transcription.

The affected subject was a Caucasian of Mediterranean origin born to consanguineous parents. He had prenatal diagnosis of myelomeningocele (lumbosacral) with hydrocephalus, and at birth the weight and head circumference were on the 50th centile. A shortened thorax was noted, as well as an ectopic stenotic anus and talipes. Radiologically, the spine was short with multiple contiguous SDV. A cerebral CT scan showed a Chiari II malformation. Growth parameters at 4 years of age were 3rd centile for height and weight, but 98th centile for head circumference. Having determined from DNA sequencing that no mutation was found in *DLL3*, *MESP2,* or *LFNG*, analysis of homozygosity of descent using Affymetrix

GeneChip® Human Mapping 250 K Sty Array revealed a single candidate region in the proband – a 10.1 Mb on chromosome 17p13. The interval included two candidate genes, namely *HES7* and *DVL2* (*DISHEVELLED2*). On sequencing both genes, homozygosity for a missense mutation was found in *HES7*, c.73C > T (R25W, Sparrow et al. 2008). Both parents were heterozygous for the mutant allele, one unaffected sibling was heterozygous, and the other unaffected sibling homozygous normal. The mutation was not found in 110 ethnically matched controls (220 chromosomes).

HES family proteins repress transcription through two distinct mechanisms. These proteins bind directly to DNA via an N-box (CACNAG) using a basic region immediately N-terminal to the helix–loop–helix domain (which is involved in homo- and heterodimerization with other bHLH family members). Corepressors are then recruited to the promoter via interaction with a WRPW C- terminal motif. They can also form heterodimers with the bHLH protein E47, thus preventing it (and other bHLH proteins that normally heterodimerize with E47 such as MyoD) from binding to E-boxes (CANNTG) and activating transcription.

Functional analysis was undertaken using a beta-actin reporter with upstream N-boxes or E-boxes, and comparisons of transcriptional repression were made between the wild-type mouse Hes7 with that of the R25W mutant Hes7. Levels of transcription of the mutant allele were significantly increased, suggesting it lacked normal repression activity. The evidence therefore supports this mutation as causative of the phenotype, and this is designated SCD type 4, or *HES7*-associated SCD. The actual phenotype (Fig. 5.6) bears a close similarity to STD with posterior fusion of the ribs that fan out in a "crab-like" manner. The abnormal segmentation of the vertebral bodies closely resembles that seen in *DLL3*-associated SCD (type1), though

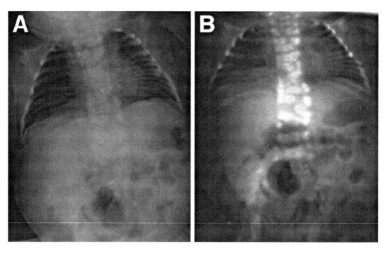

Fig. 5.6 The radiological phenotype of SCD type 4 due to mutated *HES7*. The pattern is similar to STD, with posterior fusion of the ribs that fan out in a "crab-like" manner. Also, the vertebral pedicles are relatively prominent compared to SCD type 1 – the "tramline sign". From Sparrow et al. 2006

the vertebral pedicles are also evident, as in STD – the "tramline sign." Until more cases of *HES7*-associated SCD are reported, it is unclear whether the neural tube defect (and ectopic anus) in this patient was directly linked to mutated *HES7*.

Mesoderm Posterior 2 (*MESP2*) and Spondylothoracic Dysostosis (STD)

Spondylothoracic dysostosis (STD), despite showing similarities to autosomal recessive SCD, has distinctive phenotypic features and is therefore separately designated. As mentioned, the term "spondylothoracic dysostosis/dysplasia" was first proposed by Moseley and Bonforte (1969), the thoracic vertebrae are usually severely "telescoped" together, and consequently infant mortality is approximately 50% due to respiratory insufficiency. Survival may depend on the availability of expert respiratory and intensive care. The clinical aspects of STD have been documented in detail (Cornier et al. 2004). The radiologic findings of STD that distinguish it from SCD are (1) more severe shortening of the spine (with all vertebral segments affected), especially the thoracic spine, leading to impaired respiratory function in infancy; (2) rib fusions typically occurring posteriorly at the costovertebral origins, where the spinal shortening is most severe. The ribs usually appear straight and neatly aligned without points of fusion along their length (in contrast to the appearance in SCD). On antero-posterior X-ray the ribs characteristically "fan out" from their costovertebral origins in a "crab-like" fashion (Fig. 5.7); and (3) early radiographic prominence of the vertebral pedicles, a feature described as the "tramline sign" (Turnpenny et al. 2007). In SCD, in fetal life and early childhood, there is a significant lack of ossification of the vertebral pedicles.

STD occurs most frequently in Puerto Ricans, presumably due to the *MESP2* founder mutation, p.GluE103X (Cornier et al. 2008). In this population, the phenotype has been known as Jarcho–Levin syndrome (Jarcho and Levin 1938). However,

Fig. 5.7 The radiological phenotype of STD due to mutated *MESP2*. Severe telescoping of the spine with all vertebral segments affected, and the ribs fan out from their posterior origins with little, if any, intercostal fusions, giving a "crab-like" appearance. The early ossification of the vertebral pedicles (in contrast to that seen in SCD) has prompted the suggestion that this be termed the "tramline sign"

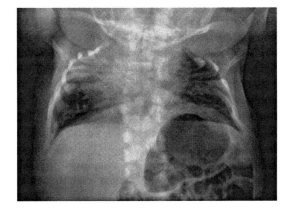

the patients originally reported by Jarcho and Levin were not Puerto Rican (they were described as 'colored') and the phenotype was closer to SCD type 2.

To date most individuals reported with STD have had nonsense mutations in exon 1 of *MESP2*, which are predicted to result in nonsense mediated decay; however, several affected individuals are heterozygous for a nonsense mutation and a missense mutation (Cornier et al. 2008).

A New Classification of SDV

Progress in identifying the genetic basis of different forms of SDV is hindered by the broad diversity of phenotypes, their individual rarity, and the lack of a systematic approach to classification and stratification. However, a collaborative multicenter, multidisciplinary, approach could facilitate the best opportunity to both elucidate the causes of SDV as well as conduct clinical studies focussed on the natural history and treatment of these disorders. To these ends a consortium, the International Consortium for Vertebral Anomalies and Scoliosis (ICVAS), was formed in November 2005, comprising developmental biologists, radiologists, clinical geneticists, and orthopaedic surgeons. The group has invested time in developing a new classification system in order to facilitate consistent phenotype delineation, one that is sufficiently robust to inform phenotype–genotype correlation studies. The system can use ontology applicable to both humans and animal models, and is presented diagrammatically in Figs. 5.8 and 5.9. It is essentially a descriptive system, based on the radiology of the spine, and accommodates current knowledge of the Mendelian forms of SCD as well as well-recognized syndromes where SDV is a feature.

In this scheme, we recommend that terms SCD and STD are reserved for specific phenotypes (Table 5.2). Strictly, these are dysostoses, not dysplasias, because they are due to errors of segmentation or formation early in morphogenesis, rather than an ongoing abnormality of chondro-osseous tissues during prenatal and postnatal life. In SCD or STD the thoracic cage must have a *general* appearance of symmetry, which may still be evident despite a small difference in rib number on each side. We propose at least 10 contiguous vertebral bodies should be affected because, to date, in all known Mendelian cases of SCD, virtually all vertebrae show evidence of abnormal segmentation and involvement is contiguous rather than multiregional. In SCD, there are usually points of fusion of the ribs along their length, while in STD the ribs are fused *posteriorly*, and they fan out laterally without points of fusion along their length ("crab-like" appearance). In early life, multiple rounded hemivertebrae ("pebble beach sign," Turnpenny et al. 2003) characterizes SCD type 1 (*DLL3* gene) and the vertebral pedicles are poorly visualized (Fig. 5.3). In *MESP2*-associated SCD and STD, again in early life, by contrast the vertebral pedicles are well formed and prominent (Fig. 5.6), for which we suggest the term "tramline sign." In cases where at least 10 vertebral segments are affected but non-contiguously, we suggest this is designated a "multiregional" form of M-SDV rather

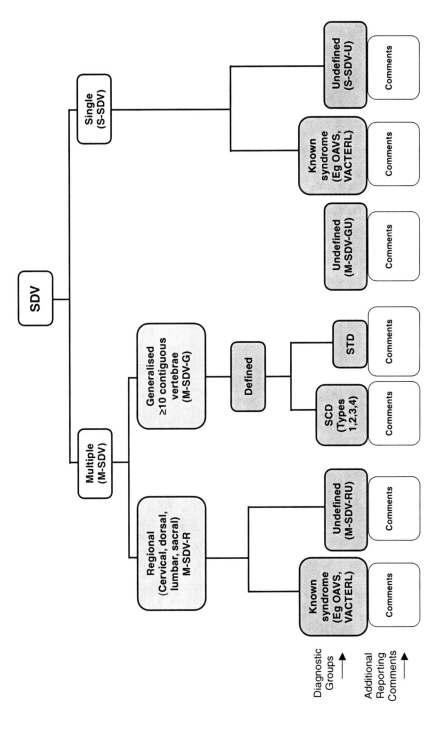

Fig. 5.8 ICVAS clinical classification algorithm. SDV: Segmentation defect(s) of the vertebrae; M: Multiple; S: Single; R: Regional; G: Generalized; U: Undefined

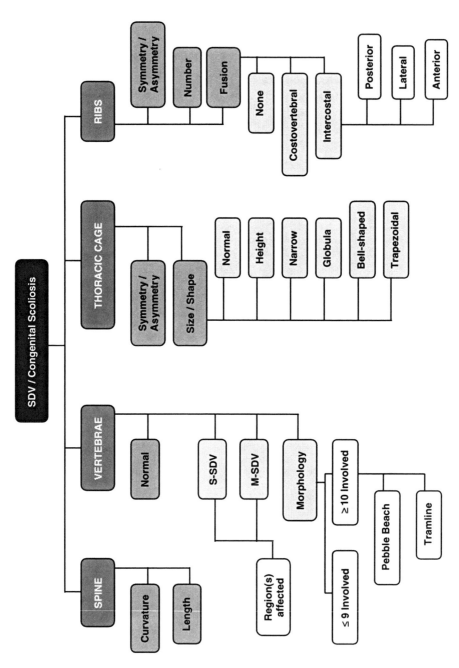

Fig. 5.9 ICVAS classification algorithm for research. A more detailed, systematic analysis of radiographic anatomical features

than "generalized." This group of phenotypes appears to be very diverse and further delineation will only be possible with advances in our understanding of causation.

We aimed to produce a classification system for SDV that provides simple, uniform terminology and can be applied both to man and to animal models. The role played by correlation of a detailed clinical examination with radiological findings has been well described (Erol et al. 2004). The system is suitable for routine clinical reporting (Fig. 5.8), and an extended version is suitable for gathering more detailed data for research purposes (Fig. 5.9). Where appropriate, the new classification incorporates existing terminology. For example, types of spinal abnormality have been described according to the classification of Aburakawa et al. (1996) (Table 5.8).

In the new system, as far as clinical reporting is concerned, conditions fall into one of seven basic categories (Fig. 5.8). This simplification allows for uniformity between observers. For example, a case with limited thoracic spine involvement that might previously have been diagnosed as Jarcho–Levin syndrome or SCD would now be classified and reported as an "undefined regional (thoracic) M-SDV." This greatly reduces confusion that might be generated by indiscriminate use of the terms JLS or SCD.

This system is simple because it is descriptive and has potential to be reliable. It allows for a more precise characterization of the radiological phenotype by incorporating phenotypic patterns of the spine as a whole, besides formation errors of individual vertebrae. We believe a consistent use of terminology using this system will lead to improved diagnostic consensus and better stratification of patient cohorts for testing of novel gene candidates and evaluation of natural history. It

Table 5.8 Classification of vertebral segmentation abnormalities (Aburukawa et al. 1996, Takikawa et al. 2006) (modified North American classification). Note that hemivertebrae are seen in types B to F and L

Failure of formation
Type I
A. Double pedicle
B. Semi-segmented
C. Incarcerated

Type II
D. Non-incarcerated, no lateral shift
E. Non-incarcerated, plus lateral shift

Type III
F. Multiple

Type IV
G. Wedge
H. Butterfly

Failure of segmentation
I. Unilateral Bar
J. Complete block
K. Wedge (plus narrow disk)

Mixed
L. Unilateral bar plus hemivertebra
M. Unclassifiable

can be readily modified as our understanding of distinctive phenotypes, genotype–phenotype correlations, and natural histories evolves. It is a tool to aid more accurate diagnosis and research. Clinical management of the individual patient is naturally much more complex and involves much assessment and surveillance of growth, respiratory function, and spinal curvature.

Concluding Remarks

Abnormal segmentation of the spine is an important "handle" in clinical dysmorphology. The number and range of conditions and syndromes that include this malformation as a feature is very large, but so too is the range of phenotypes for which no clear diagnosis exists. Molecular genetics has enabled the identification of the cause of some phenotypes that follow Mendelian inheritance, usually through studies undertaken on affected families suitable for DNA linkage analysis. The insights and clues gained from animal studies and model systems have also been crucial to identifying candidate genes and will continue to be so. It seems likely that the monogenic causes of axial skeletal defects in man will, in due course, yield their secrets. Somitogenesis genes have justifiably been the focus of huge attention but many other genes and mechanisms have been implicated in normal axial skeletal development, and these need to be explored in cohorts of affected patients whose clinical features are stratified systematically. It is likely that a large proportion of the phenotypes seen in clinical practice, if not the majority, are not monogenic in causation but multifactorial. This large and diverse group, which collectively accounts for the most difficult medical and surgical problems, is likely to present the greatest challenge in the field of abnormal spinal segmentation.

References

Aburakawa, K., Harada, M., and Otake, S. 1996. Clinical evaluations of the treatment of congenital scoliosis. Orthop. Surg. and Trauma 39:55–62.

Aurora, P., Wallis, C.E., and Winter, R.M. 1996. The Jarcho-Levin syndrome (spondylocostal dysplasia) and complex congenital heart disease: a case report. Clin. Dysmorphol. 5:165–169.

Aymé, S. and Preus, M. 1986. Spondylocostal/spondylothoracic dysostosis: the clinical basis for prognosticating and genetic counselling. Am. J. Med. Genet. 24:599–606.

Bartsocas, C.S., Kiossoglou, K.A., Papas, C.V., et al. 1981. Spondylocostal dysostosis in South African sisters. Clin. Genet. 19:23–25.

Bessho, Y., Sakata, R., Komatsu, S., Shiota, K., Yamada, S., and Kageyama, R. 2001. Dynamic expression and essential functions of Hes7 in somite segmentation. Genes Develop. 15:2642–2647.

Bohring, A., Lewin, S.O., Reynolds, J.F. et al. 1999. Polytopic anomalies with agenesis of the lower vertebral column. Am. J. Med. Genet. 87:99–114.

Bonaime, J.L., Bonne, B., Joannard, A. et al. 1978. Le syndrome de dysostose spondylothoracique ou spondylocostale. Pédiatrie 33:173–188.

Bruckner, K., Perez, L., Clausen, H. et al. 2000. Glycosyltransferase activity of Fringe modulates Notch-Delta interactions. Nature 406:411–415.

Bulman, M.P., Kusumi, K., Frayling, T.M. et al. 2000. Mutations in the human Delta homologue, DLL3, cause axial skeletal defects in spondylocostal dysostosis. Nature Genet. 24:438–441.

Cantú, J.M., Urrusti, J., Rosales, G. et al. 1971. Evidence for autosomal recessive inheritance of costovertebral dysplasia. Clin. Genet. 2:149–154.

Casamassima, A.C., Morton, C.C., Nance, W.E. et al. 1981. Spondylocostal dysostosis associated with anal and urogenital anomalies in a Mennonite sibship. Am. J. Med. Genet. 8:117–127.

Chen, J., Moloney, D.J., and Stanley, P. 2001. Fringe modulation of Jagged1-induced Notch signaling requires the action of beta 4 galactosyltransferase-1. Proc. Natl. Acad. Sci. U. S. A 98:13716–13721.

Clarke, R., Catalan, G., Diwan, A., Kearsley, J. 1998. Heterogenetiy in Klippel-Feil syndrome: a new classification. Pediatr. Radiol. 28:967–974.

Cornier, A.S., Ramrez-Lluch, N., Arroyo, S. et al. 2000. Natural history of Jarcho-Levin syndrome (Abstract). Am. J. Hum. Genet. 67(4)(Suppl.2):56(A238).

Cornier, A.S., Ramírez, N., Arroyo, S. et al. 2004. Phenotype characterisation and natural history of spondylothoracic dysplasia syndrome: a series of 27 new cases. Am. J. Med. Genet. 128A: 120–126.

Cornier, A.S., Staehling-Hampton, K., Delventhal, K.M. et al. 2008. Mutations in the *MESP2* gene cause spondylothoracic Dysostosis/Jarcho-Levin syndrome. Am. J. Hum. Genet. 82:1334– 1341.

Correia, T. et al. 2003. Molecular genetic analysis of the glycosyltransferase Fringe in Drosophila. Proc. Natl. Acad. Sci. U. S. A. 100:6404–6409.

Crow, Y.J., Tolmie, J.L., Rippard, K. et al. 1997. Spondylocostal dysostosis associated with a 46,XX,+15,dic(6;15)(q25;q11.2) translocation. Clin. Dysmorphol. 6:347–350.

Daikha-Dahmane, F., Huten, Y., Morvan, J., Szpiro-Tapia, S., Nessmann, C., and Eydoux, P. 1998. Fetus with Casamassima-Morton-Nance syndrome and an inherited (6;9) balanced translocation. Am. J. Med. Genet. 80:514–517.

David, T.J. and Glass, A. 1983. Hereditary costovertebral dysplasia with malignant cerebral tumour. J. Med. Genet. 20:441–444.

De Grouchy, J., Mlynarski, J.C., Maroteaux, P. et al. 1963. Syndrome polydysspondylique par translocation 14–15 et dyschondrostéose chez un même sujet. Ségrégation familiale. C. R. Acad. Sci. [D] Paris 256:1614–1616.

Delgoffe, C., Hoeffel, J.C., Worms, A.M. et al. 1982. Dysostoses spondylocostales et cardiopathies congénitales. Ann. Pédiat. 29:135–139.

Devos, E.A., Leroy, J.G., Braeckman, J.J. et al. 1978. Spondylocostal dysostosis and urinary tract anomaly: definition and review of an entity. Eur. J. Paed. 128:7–15.

Dowton, S.B., Hing, A.V., Sheen-Kaniecki, V. et al. 1997. Chromosome 18q22.2-qter deletion and a congenital anomaly syndrome with multiple vertebral segmentation defects. J. Med. Genet. 34:414–417.

Dunwoodie, S.L., Henrique, D., Harrison, S.M. et al. 1997. Mouse Dll3: a novel divergent Delta gene which may complement the function of other Delta homologues during early pattern formation in the mouse embryo. Development 124:3065–3076.

Dunwoodie, S.L., Clements, M., Sparrow, D.B. et al. 2002. Axial skeletal defects caused by mutation in the spondylocostal dysplasia/pudgy gene Dll3 are associated with disruption of the segmentation clock within the presomitic mesoderm. Development 129:1795–1806.

Eliyahu, S., Weiner, E., Lahav, D. et al. 1997. Early sonographic diagnosis of Jarcho-Levin syndrome: a prospective screening program in one family. Ultrasound Obstet. Gynecol. 9:314–318.

Eller, J.L. and Morton, J.M. 1976. Bizarre deformities in offspring of user of lysergic acid diethylamide. New Eng. J. Med. 283:395–397.

Erol, B., Tracy, M.R., Dormans, J.P., Zackai, E.H., Maisenbacher, M.K., O'Brien, M.L., Turnpenny, P.D., and Kusumi, K. 2004. Congenital scoliosis and vertebral malformations: characterization of segmental defects for genetic analysis. J. Pediatr. Orthop. 24:674–682.

Feil, A. 1919. L'absence et la diminution des vertebres cervicales (etude clinique et pathologique); le syndrome de recution numeriique cervicale. Thesis de Paris.

Floor, E., De Jong, R.O., Fryns, J.P. et al. 1989. Spondylocostal dysostosis: an example of autosomal dominant inheritance in a large family. Clin. Genet. 36:236–241.

Geffers, I., Serth, K., Chapman, G. et al. 2007. Divergent functions and distinct localization of the Notch ligands DLL1 and DLL3 in vivo. J. Cell Biol. 178(3):465–476.

Gellis, S.S. and Feingold, M. 1976. Picture of the month: spondylothoracic dysplasia. Am. J. Dis. Child 130:513–514.

Giacoia, G.P. and Say, B. 1991. Spondylocostal dysplasia and neural tube defects. J. Med. Genet. 28:51–53.

Giampietro, P.F., Raggio, C.L., and Blank, R.D. 1999. Synteny-defined candidate genes for congenital and idiopathic scoliosis. Am. J. Med. Genet. 83:164–177.

Giampietro, P.F., Raggio, C.L., Reynolds, C. et al. 2006. DLL3 as a candidate gene for vertebral malformations. Am. J. Med. Genet. A 140(22):2447–2453.

Gridley, T. 1997. Notch signaling in vertebrate development and disease. Mol. Cell. Neurosci. 9:103–108.

Gridley, T. 2003. Notch signaling and inherited disease syndromes. Hum. Mol. Genet. 12(Rev 1):R9–R13.

Grüneberg, H. 1961. Genetical studies on the skeleton of the mouse. Genet. Res. Camb. 2:384–393.

Haines, N. and Irvine, K.D. 2003. Glycosylation regulates Notch signalling. Nat. Rev. Mol. Cell. Biol. 4:786–797.

Harper, J.A., Yuan, J.S., Tan, J.B. et al. 2003. Notch signaling in development and disease. Clin. Genet. 64:461–472.

Herold, H.Z., Edlitz, M., and Barochin, A. 1988. Spondylothoracic dysplasia. Spine 13:478–481.

Jarcho, S. and Levin, P.M. 1938. Hereditary malformation of the vertebral bodies. Bull. Johns Hopkins Hosp. 62:216–226.

Joutel, A., and Tournier-Lasserve, E. 1998. Notch signalling and human disease. Cell. Dev. Biol. 9:619–625.

Karnes, P.S., Day, D., Barry, S.A. et al. 1991. Jarcho-Levin syndrome: four new cases and classification of subtypes. Am. J. Med. Genet. 40:264–270.

Kozlowski, K. 1984. Spondylo-costal dysplasia. Fortschr. Röntgenstr. 140:204–209.

Kusumi, K., Sun, E.S., Kerrebrock, A.W. et al. 1998. The mouse pudgy mutation disrupts Delta homologue Dll3 and initiation of early somite boundaries. Nat. Genet. 19:274–278.

Kusumi, K., Dunwoodie, S.L., Krumlauf, R. 2001. Dynamic expression patterns of the pudgy/spondylocostal dysostosis gene Dll3 in the developing nervous system. Mech. Dev. 100:141–144.

Kusumi, K., Mimoto, M.S., Covello K. et al. 2004. Dll3 pudgy mutation differentially disrupts dynamic expression of somite genes. Genesis 39:115–121.

Ladi, E., Nichols, J.T., Ge, W. et al. 2005. The divergent DSL ligand Dll3 does not activate Notch signaling but cell autonomously attenuates signaling induced by other DSL ligands. J. Cell Biol. 170:983–992.

Lavy, N.W., Palmer, C.G., and Merritt, A.D. 1966. A syndrome of bizarre vertebral anomalies. J. Pediatr. 69:1121–1125.

Lendon, R.G., Wynne-Davies, R., and Lendon, M. 1981. Are congenital vertebral anomalies and spina bifida cystica aetiologically related? J. Med. Genet. 18:424–427.

Lin, A.E. and Harster, G.A. 1993. Another case of spondylocostal dyplasia and severe anomalies (letter). Am. J. Med. Genet. 46:476–477.

McCall, C.P., Hudgins, L., Cloutier, M. et al. 1994. Jarcho-Levin syndrome: unusual survival in a classical case. Am. J. Med. Genet. 49:328–332.

McGrew, M.J., Pourquié, O. 1998. Somitogenesis: segmenting a vertebrate. Curr. Opinion Genet. Dev. 8:487–493.

McLennan, J.E. 1976. Rib anomalies in myelodysplasia. Biol. Neonate 29:129–141.

McMaster, M.J. and Singh, H. 1999. Natural history of congenital kyphosis and congenital kyphoscoliosis. A study of one hundred and twelve patients. J. Bone. Joint Surg. Am. 81:1367–1383.

Maisenbacher, M.K., Han, J.S., O'Brien, M.L. et al. 2005. Molecular analysis of congenital scoliosis: a candidate gene approach. Hum. Genet. 116(5):416–419.

Martínez-Frías, M.L., Bermejo, E., Paisán, L. et al. 1994. Severe spondylocostal dysostosis associated with other congenital anomalies: a clinical/epidemiological analysis and description of ten cases from the Spanish Registry. Am. J. Med. Genet. 51:203–212.

Martínez-Frías, M.L. 1996. Multiple vertebral segmentation defects and rib anomalies. Am. J. Med. Genet. 66:91.

Massari, M.E. and Murre, C. 2000. Helix-loop-helix proteins: regulators of transcription in eucaryotic organisms. Mol. Cell Biol. 20:429–440.

Moloney, D.J. et al. 2000. Fringe is a glycosyltransferase that modifies Notch. Nature 406: 369–375.

Mortier, G.R., Lachman, R.S., Bocian, M. et al. 1996. Multiple vertebral segmentation defects: analysis of 26 new patients and review of the literature. Am. J. Med. Genet. 61:310–319.

Moseley, J.E. and Bonforte, R.J. 1969. Spondylothoracic dysplasia – a syndrome of congenital anomalies. Am. J. Roentgenol. 106:166–169.

Murr, M.M., Waziri, M.H., Schelper, R.L. et al. 1992. Case of multiple vertebral anomalies, cloacal dysgenesis, and other anomalies presenting prenatally as cystic kidneys. Am. J. Med. Genet. 42:761–765.

Naik, P.R., Lendon, R.G., and Barson, A.J. 1978. A radiological study of vertebral and rib malformations in children with myelomeningocele. Clin. Radiol. 29:427–430.

Nakano, S., Okuno, T., Hojo, H. et al. 1977. 18p- syndrome associated with hemivertebrae, fused ribs and micropenis. Jpn. J. Hum. Genet. 22:27–32.

Norum, R.A. and McKusick, V.A. 1969. Costovertebral anomalies with apparent recessive inheritance. Birth Defects O.A.S. 18:326–329.

Novak, R.W. and Robinson, H.B. 1994. Coincident DiGeorge anomaly and renal agenesis and its relation to maternal diabetes. Am. J. Med. Genet. 50:311–312.

Offiah, A.C. and Hall, C.M. 2003. Radiological diagnosis of the constitutional disorders of bone. As easy as A, B, C? Pediatr. Radiol. 33:153–161.

Ohzeki, T., Shiraishi, M., Matsumoto, Y. et al. 1990. Sporadic occurrence of spondylocostal dysplasia and mesocardia in a Japanese girl. Am. J. Med. Genet. 37:427–428

Pérez-Comas, A., and García-Castro, J.M. 1974. Occipito-facial-cervico-thoracic-abdomino-digital dysplasia: Jarcho-Levin syndrome of vertebral anomalies. J. Pediatr. 85:388–391.

Pochaczevsky, R., Ratner, H., Perles, D. et al. 1971. Spondylothoracic dysplasia. Radiology 98: 53–58.

Poor, M.A., Alberti, A., Griscom, T. et al. 1983. Nonskeletal malformations in one of three siblings with Jarcho-Levin syndrome of vertebral anomalies. J. Pediatr. 103:270–272.

Pourquié O. 1999. Notch around the clock. Curr. Opinion Genet. Dev. 9:559–565.

Pourquié, O., Kusumi, K. 2001. When body segmentation goes wrong. Clin. Genet. 60:409–416.

Purkiss, S.B., Driscoll, B., Cole, W.G. et al. 2002. Idiopathic scoliosis in families of children with congenital scoliosis. Clin. Orthop. Relat. Res. 401:27–31.

Rastogi, D., Rosenzweig, E.B., and Koumbourlis, A. 2002. Pulmonary hypertension in Jarcho-Levin syndrome. Am. J. Med. Genet. 107:250–252.

Reyes, M.C., Morales, A., Harris, V. et al. 1989. Neural defects in Jarcho-Levin syndrome. J. Child Neurol. 4:51–54.

Riccardi, V.M. 1977. Trisomy 8: an international study of 70 patients. Birth Defects O.A.S. XIII(3C):171–184.

Roberts, A.P., Conner, A.N., Tolmie, J.L. et al. 1988. Spondylothoracic and spondylocostal dysostosis. J. Bone Joint Surg. 70B:123–126.

Saga, Y., Hata, N., Koseki, H. et al. 1997. Mesp2: a novel mouse gene expressed in the presegmented mesoderm and essential for segmentation initiation. Genes Dev. 11:1827–1839.

Schulman, M., Gonzalez, M.T., and Bye, M.R. 1993. Airway abnormalities in Jarcho-Levin syndrome: a report of two cases. J. Med. Genet. 30:875–876.

Sharma, A.K. and Phadke, S.R. 1994. Another case of spondylocostal dysplasia and severe anomalies: a diagnostic and counseling dilemma. Am. J. Med. Genet. 50:383–384.

Simpson, J.M., Cook, A., Fagg, N.L.K. et al. 1995. Congenital heart disease in spondylothoracic dysostosis: two familial cases. J. Med. Genet. 32:633–635.

Solomon, L., Jimenez, B., and Reiner, L. 1978. Spondylothoracic dysostosis. Arch. Pathol. Lab. Med. 102:201–205.

Sparrow, D.B., Chapman, G., Wouters, M.A. et al. 2006. Mutation of the LUNATIC FRINGE gene in humans causes spondylocostal dysostosis with a severe vertebral phenotype. Am. J. Hum. Genet. 78:28–37.

Sparrow, D.B., Guillén-Navarro, E., Fatkin, D., and Dunwoodie, S.L. 2008. Mutation of Hairy-and-Enhancer-of-Split-7 in humans causes spondylocostal dysostosis. Hum. Mol. Genet. 17(23):3761–3766.

Superti-Furga, A., and Unger, S. 2007. Nosology and classification of genetic skeletal disorders: 2006 revision. Am. J. Med. Genet. 143:1–18.

Takahashi, Y., Koizumi, K., Takagi, A. et al. 2000. Mesp2 initiates somite segmentation through the Notch signalling pathway. Nat. Genet. 25:390–396.

Takahashi, Y., Inoue, T., Gossler, A. et al. 2003. Feedback loops comprising Dll1, Dll3 and Mesp2, and differential involvement of Psen1 are essential for rostrocaudal patterning of somites. Development 130:4259–4268.

Takikawa, K., Haga, N., Maruyama, T., Nakatomi, A., Kondoh, T., Makita, Y., Hata, A., Kawabata, H., and Ikegawa, S. 2006. Spine and rib abnormalities and stature in spondylocostal dysostosis. Spine 31:E192–E197.

Thomsen, M., Schneider, U., Weber, M., Johannisson, R., Niethard, F. 1997. Scoliosis and congenital anomalies associated with Klippel-Feil syndrome types I–III. Spine 22:396–401.

Tolmie, J.L., Whittle, M.J., McNay, M.B. et al. 1987. Second trimester prenatal diagnosis of the Jarcho-Levin syndrome. Prenat. Diagn. 7:129–134.

Trindade, C.E.P., and de Nóbrega, F.J. 1977. Spondylothoracic dysplasia in two siblings. Clin. Pediatr. 16:1097–1099.

Turnpenny, P.D., Thwaites, R.J., and Boulos, F.N. 1991. Evidence for variable gene expression in a large inbred kindred with autosomal recessive spondylocostal dysostosis. J. Med. Genet. 28:27–33.

Turnpenny, P.D., Bulman, M.P., Frayling, T.M. et al. 1999. A gene for autosomal recessive spondylocostal dysostosis maps to 19q13.1-q13.3. Am. J. Hum. Genet. 65:175–182.

Turnpenny, P.D., Whittock, N.V., Duncan, J. et al. 2003. Novel mutations in DLL3, a somitogenesis gene encoding a ligand for the Notch signalling pathway, cause a consistent pattern of abnormal vertebral segmentation in spondylocostal dysostosis. J. Med. Genet. 40:333–339.

Turnpenny, P.D., Alman, B., Cornier, A.S., Giampietro, P.F., Offiah, A., Tassy, O., Pourquié, O., Kusumi, K., and Dunwoodie, S. 2007. Abnormal vertebral segmentation and the Notch signaling pathway in man. Dev. Dyn. 236:1456–1474.

Whittock, N.V., Ellard, S., Duncan, J. et al. 2004a. Pseudo-dominant inheritance of spondylocostal dysostosis type 1 caused by two familial delta-like 3 mutations. Clin. Genet. 66:67–72.

Whittock, N.V., Sparrow, D.B., Wouters, M.A. et al. 2004b. Mutated MESP2 causes spondylocostal dysostosis in humans. Am. J. Hum. Genet. 74:1249–1254.

Wynne-Davies, R. 1975. Congenital vertebral anomalies: aetiology and relationship to spina bifida cystica. J. Med. Genet. 12:280–288.

Chapter 6
Spondylothoracic Dysostosis in Puerto Rico

Alberto Santiago Cornier

Introduction

Puerto Rico is the smallest of the four Greater Antilles islands, which also includes Cuba, Hispaniola, and Jamaica, localized in the Caribbean Sea. Puerto Rico was discovered in 1493 by Christopher Columbus and became a Spanish territory until the Spanish–American War of 1898, when it became a US territory. The particular history of our country provides us with a specific social, cultural, and ethnic situation. We are the only Latin-American country whose people have US citizenship. The two official languages of Puerto Rico are Spanish and English. The island population is 3.9 million (U.S. Census Bureau 2000, http://www.census.gov) and taking into consideration that the island is only 35 miles wide and 100 miles long, this makes us a densely populated country with 1,115.8 people per square mile.

The ethnic background of the Puerto Rican population is almost entirely the product of the mixture of Spaniards, Africans, and Taíno Indians. There is also some contribution to the population from other southern European countries. Although there is no specific genetic isolation in Puerto Rico, there is high degree of consanguinity, particularly in the central mountain region of the island. Like the rest of the Caribbean islands, Puerto Rico had limited economic resources and prominent poverty until the 1960s, when an industrial revolution significantly improved the lives of its citizens. This economic data is relevant since this is one of the reasons people tend to marry within the same family or with neighbors in the same village. When a mutation segregates within the genetic pool of such populations with nonrandom mate selection, an increase in prevalence of such mutation may occur rapidly, producing a high incidence of otherwise rare genetic disorders.

The human genome contains 3.1647 billion chemical nucleotide bases (A, C, T, and G). Approximately 99.9% of these nucleotide sequences are the same in all people. This leaves about 2 million nucleotides (0.1%) that are unique from person

A.S. Cornier (✉)
Department of Molecular Medicine, Hospital de la Concepción, San Germán, Puerto Rico; Ponce School of Medicine, Ponce, Puerto Rico
e-mail: scornier@hospitalconcepcion.org

K. Kusumi, S.L. Dunwoodie (eds.), *The Genetics and Development of Scoliosis*, DOI 10.1007/978-1-4419-1406-4_6, © Springer Science+Business Media, LLC 2010

to person. These sequence differences account, in part, for why we look, think, and behave differently, and some of these nucleotide variations are dramatically different among the various population groups. Genetic admixture represents the measure of the proportion of African, European, and Amerindian ancestral genetic makeup of the individuals within a specific population and may prove useful to learn more about its relation to disease. Genetic admixture is receiving greater attention because of its implications for gene mapping, its implications to personalized medicine, and its role in understanding health disparities.

The unique history and population characteristics of Puerto Rico, with genetic admixture and geographic isolation, combine to provide an opportunity for investigating the etiology of genetic congenital disorders such as vertebral abnormalities. The population of Puerto Rico experienced a genetic bottleneck soon after Western contact when there may have been a severely limited founder population. Subsequent episodic immigration of Europeans and Africans has resulted in a current population of nearly 4 million. Mitochondrial analysis of maternal ancestry points to a mixed population of 61.3% Amerindian (including Taíno), 27% sub-Saharan African, and 11.5% West Eurasian derivation (Martínez-Cruzado et al. 2005).

Spondylothoracic Dysostosis and Segmentation Defects of the Vertebrae

Congenital segmentation defects of the vertebrae in humans are present in a wide variety of rare but well-characterized disorders, as well as many diverse and poorly understood phenotypic patterns (Turnpenny et al. 2007). In some cases, patients with congenital scoliosis and chest wall abnormalities present a major medical challenge. Jarcho and Levin (1938) described a Puerto Rican family whose two children presented with a shortened trunk with abnormal segmentation throughout the vertebral column and irregularly aligned ribs, with normal long bones and skull. Since then, the authors' names have frequently been used eponymously (and often inappropriately) for almost any form of costovertebral malformation.

Spondylothoracic dysostosis (STD), described in detail below, has been well characterized as an autosomal recessive syndrome with high prevalence in the Puerto Rican population, and the name "Jarcho–Levin syndrome" has been used to describe a variety of clinical phenotypes consisting of short-trunk dwarfism associated with rib and vertebral anomalies. This admixture of phenotypes under Jarcho–Levin syndrome has allowed some confusion in terms of phenotype, prognosis, and mortality. Recent publications have provided more insight into the clinical and molecular diagnosis, prognosis, and management of patients with these phenotypes (reviewed in Chapter 5). Based on these findings, two groups of distinct phenotypes have been determined. These are spondylo*thoracic* dysostosis (STD) and spondylo*costal* dysostosis (SCD). Both clinical entities are present in Puerto Rican population but spondylothoracic dysostosis predominates, comprising 49% of the STD cases reported in the medical literature (Cornier et al. 2004).

Segmentation defects of the vertebrae (SDV) represent an important health issue due to the resulting scoliosis/kyphosis with or without pain, limitation of movement, and respiratory complications. A vertebral malformation may be an isolated finding as well as a part of a conundrum of other morphological and clinical manifestations.

Spondylothoracic Dysostosis in the Puerto Rican population

Spondylothoracic dysostosis (MIM #277300) is a rare pleiotropic genetic disorder with autosomal recessive inheritance first described by Jarcho and Levin (1938). The term dysostosis refers to the intrinsic pathology of this disease that is strictly related to the embryologic mesoderm that forms the vertebrae. Typical findings include segmentation and formation defects throughout cervical, thoracic, and lumbar spine, such as hemivertebrae, block vertebrae, and unsegmented bars, with fusion of all the ribs at the costovertebral junction (Figs. 6.1 and 6.2). In addition to the multiple skeletal anomalies, short and rigid neck, short thorax, protuberant abdomen, inguinal and umbilical hernias, urinary tract abnormalities, and disproportionate dwarfism have been described. As mentioned before in this chapter, many names has been used to describe the phenotype for several types of short-trunk dwarfism characterized by multiple segmentation defects of the vertebral bodies and ribs, including spondylocostal dysplasia (Rimoin et al. 1968), spondylocostal dysostosis (Casamassima et al. 1981, Franceschini et al. 1974, Ayme and Preus 1986, Floor et al. 1989), spondylothoracic dysplasia (Moseley and Bonforte 1969, Pochaczevsky et al. 1971, Solomon et al. 1978, Herold et al. 1988), and costovertebral dysplasia (Cantu et al. 1971).

We have been able to study spondylothoracic dysostosis both clinically and molecularly. STD is a relatively common disorder in Puerto Rico, presenting a unique opportunity to study the natural history and genetics of important medical

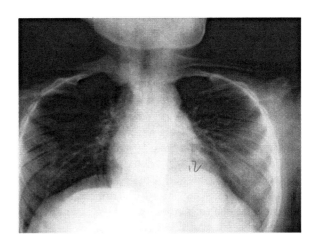

Fig. 6.1 AP X-ray view demonstrating the symmetrical and bilateral fusion of the ribs at the costovertebral junction (posteriorly) fanning out anterior-laterally

Fig. 6.2 AP X-ray showing
the multiple vertebral
formations and segmentation
defects in patients with STD

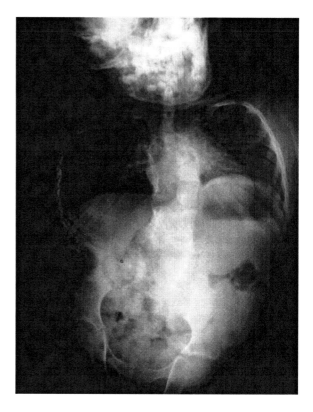

issues such as segmentation defects of the vertebrae, scoliosis, thoracic insufficiency, diaphragm dynamics, and morbidity and mortality of this disease. We
have followed a series of 38 STD Puerto Rican patients, which has allowed us to
document radiological findings, pulmonary function measurements, the molecular
etiology, and the natural history of the disorder.

Clinical Features

We have been studying patients with STD of all ages including the natural progression of those whom we evaluated as newborns and are now in their adolescence.
Characterization of childhood, adolescence, and adulthood of these patients has
been possible due to the long-term clinical follow-up.

Clinical features of STD include protuberant occipital in the newborn period that
converted to a flat occiput with aging, giving the brachycephalic appearance to the
head. Posterior hairline insertion is low in all patients. Inner canthal, interpupillary,
and outer canthal distances are normal. The nasal bridge is prominent in 33% of
the patients. The philtrum is normal in length and shape and the palate is high in

Fig. 6.3 A 12 year-old boy
with STD. Note the short
neck and anterior protrusion
of the thorax

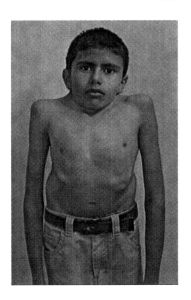

75% of patients (Cornier et al. 2004). There are two prominent clinical features
that are present in all patients and constitute the STD stigmata: these are the short
and rigid nonfunctional neck. Neck length ranges between 1.0 cm in the newborn
period and 4.7 cm in adulthood (Figs. 6.3 and 6.4). The short stature is due to a
short-trunk dwarfism. Average height percentile in male patients with STD represents 1.27th of the expected for age and sex and 3.1th in females. Due to the
posterior bilateral fusion of the ribs at the costovertebral junction, the chest opens

Fig. 6.4 An 18-month-old
baby boy with STD. Short
neck and spine are noted
particularly when compared
to the extremities

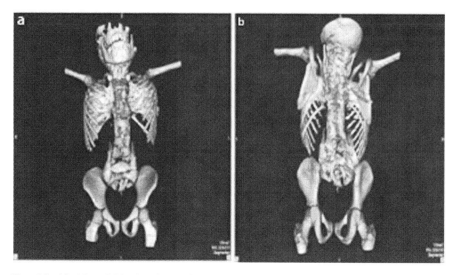

Fig. 6.5 AP (**a**) and PA (**b**) views of computer tomography scans of the whole spine with three-dimensional reformation using Xvision EX Toshiba (Japan) 2 mm cuts. All ribs were symmetrically fused at the costovertebral junction, giving the appearance of a common origin for all the ribs at the shortened thoracic spine. Photos courtesy of Dr. Norman Ramirez published in J. Bone Joint Surg. Am. 2007. 89:2663–2675

toward the anterior aspect of the thorax giving them the fanning out or "crab-like" appearance to the chest. Average chest circumference percentile measured at the nipple level was 30.85th when compared with the same age and sex normal population (Fig. 6.5a, b). Chest circumference diameter as well as spine length differences among the patients who survived versus those who did not has briefly been described (Cornier et al. 2008) and will be further discussed later on in this chapter when a phenotype/genotype analysis of STD will be discussed.

There is no significant presence of heart anomalies in patients with STD. The most common congenital heart anomaly is atrial septum defects in less than 5% of cases (Cornier et al. 2004). These atrial septum defects were all closed by age 10 years. No intrinsic lung anomaly has been described in these patients except for the hypoplastic lung syndrome at birth, secondary to the significant thoracic restriction redounding in incapacity to achieve normal size. This is of particular importance at birth since approximately 60% of babies with STD will present some type of respiratory distress requiring medical intervention. As the patients grow older the lungs tend to expand into the abdominal cavity (Fig. 6.6). Their abdominal examination is negative for enlargement of the organs (liver or spleen). The vast majority of patients with STD will develop inguinal hernias. Approximately 90% of patients do develop hernias of which up to 75% of them are bilateral in presentation. These hernias are the product of an increase in the abdominal cavity pressure due to the skeletal anomalies of the thorax and the physiological dynamics where the use of the diaphragm muscle is excessive. Due to the extensive ribs fusion in the entire

Fig. 6.6 PA view of
computer tomography scans
of the whole spine with
three-dimensional
reformation using Xvision
EX Toshiba (Japan) 2 mm
cuts. Note the close anatomic
relation of the scapulae and
the iliac crests. These
anatomic changes allow the
lungs to expand into the
abdominal cavity

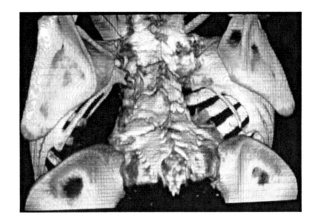

posterior and sometimes lateral chest wall the intercostal muscles are incapacitated to perform their function in the expansion of the chest. Approximately 15% of STD patients do present with umbilical hernias as well. Interestingly, surgical procedures to correct the hernias are well tolerated from respiratory and surgical points of view even if it is corrected as early as 8 months of age. There is no associated mortality reported in surgical procedures in these patients. Other congenital anomalies reported occurring with STD include soft cleft palate (less than 1%), talipes equinovarus (1.2%), unilateral glenoid agenesis (0.5%), and double urinary collector system (1%).

Since one of the major objectives of this book is to put in context scoliosis in relation to clinical and genetic perspectives, I would like to specifically comment on the presentation of scoliosis on STD patients. Scoliosis is not of frequent presentation in STD. The reason is that these patients present bilateral and symmetrical fusion of the ribs at the costovertebral junction. So forth there is little tethering effect in the spine (Fig. 6.7). The growth of the spine is intimately related to chest and lung development, with the most rapid growth occurring from birth to 5 years of age and then again in puberty (Dimeglio 1993, 2005). The greatest challenge with early onset scoliosis is that the available potential for spine and chest growth can lead to severe spinal deformity predisposing these patients to poor quality of life and thoracic insufficiency (Ramirez et al. 2009). Therefore patients with significant vertebral segmentation disorders, where there is an unbalanced tethering effect on the spine, may develop significant scoliosis. Patients with STD who present a congenital anomaly known as a bone bar may develop scoliosis. Approximately 20% of patients with STD develop spine rotation anomalies causing the spine to be rotated in its own axis and with minimal to mild development of scoliosis (Cornier et al. 2004). In contrast to STD, patients with spondylocostal dysostosis (SCD) frequently present with early onset scoliosis due to the vertebral segmentation defects with an asymmetrical tethering effect as a consequence of an asymmetrical ribs fusion.

Fig. 6.7 AP view of
computer tomography scans
of the whole spine with
three-dimensional
reformation using Xvision
EX Toshiba (Japan) 2 mm
cuts. Note the significant
fusion of the entire posterior
chest wall and the vertebral
segmentation defects

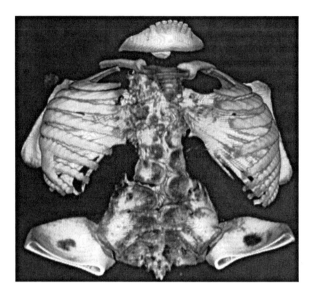

Scoliosis in patients with SCD will progress and will need surgical intervention such as the vertical expandable prosthetic titanium ribs.

Neurologically, patients with STD have no documented anomalies. The deep tendon reflexes as well as muscle tone and strength are normal. Development of neurocognitive functions is also generally normal. Patients may present mild delay in some of their milestones such as seating, crawling, and walking, which are likely to be secondary to the lack of neck motion and the disproportionate extremity to spinal length. Adults with STD have achieved professional goals, including doctoral and postdoctoral degrees.

Raising a Child with Spondylothoracic Dysostosis

I have always been fascinated by the nature of human beings and the capacity of coping that many people with genetic disorders have. For this reason and taking advantage of the fantastic research work done by Dr. Jose Acevedo Ramos in coping mechanisms in patients with STD, I will share some of his findings in this chapter.

A descriptive, cross-sectional study was conducted to explore the mood and coping strategies of Hispanic parents raising a child with spondylothoracic dysostosis. The final number of subjects in the study was 100 and included 50 mothers and 50 fathers; 37 were parents raising the same child. In addition to collecting demographic data, two measures were used to interview participants after obtaining informed consent: the Profile of Mood States (POMS), which identifies 65 mood characteristics and 6 mood states, and the Coping Health Inventory for Parents (CHIP), which lists coping strategies and identifies three coping patterns (Acevedo 2009).

The POMS results showed that *vigor/activity* was the most prevalent mood state and that *carefree* was the predominant mood characteristic. Interestingly, *carefree* is a positive adjective. Other mood characteristics present in 20–70% of the samples were *relaxed, anxious, bushed, vigorous, shaky, exhausted, panicky,* and *blue.* Two out of these additional adjectives are positive (*relaxed* and *vigorous*). These three selections seem to indicate a healthy or positive mood characteristic for many of the respondents.

This result can be explained by understanding the role that religion has in the people of Puerto Rico. Most of the people in Puerto Rico are Roman Catholic and religion is a part of everyday life in the family. Members of the family normally participate in religious events together. When events like a birth of a child affected by a particular disorder occurs in a family, often a faith-based response is shared in a family. Perhaps they view the genetic inheritance or subsequent illness as God's decision, and with this view there may be a need to accept the condition, be positive, and adjust to the event, as God will not bring them something they are not able to handle. Faith may make them feel that in some way they are special, selected by God to take care of that special child. As religious parents, they may believe they did not have any control on the birth of the child with the special needs. So being religious and having STD could alone or combined with other mood states bring parents "carefree" feeling. Feeling carefree is also supported by the low expression of guilt; only 12% of these parents reported feeling guilty compared to parents in other genetic disorders (Acevedo 2009).

Parents used over half of the possible coping strategies often. "Belief in God" scored the highest in terms of most helpful strategies and was a strategy used by 100% of participants. Similarly, "Reading about the medical condition" received the second highest score for helpfulness and was also used by 100% of participants. "Doing things with children" received the third highest score and was used by 99% of participants. A "Closer relationship with spouse" was listed as a coping strategy by 92% of parents, but was in the middle of the "most helpful" scores. "Investing myself in the children" was a strategy highly used by 98% of the parents and also was in the middle of the "most helpful" (Acevedo 2009). The coping pattern described as "Maintaining family integration, cooperation, and an optimistic definition of the situation" by the authors of CHIP was predominant among these parents. The authors label this pattern coping pattern 1. Among the coping strategies within this pattern, "Believing in God" received the most helpful averaged score (Acevedo 2009).

Perhaps depending on and strengthening family relationships is helpful in many chronic childhood diseases and this finding may be useful to clinicians. Also, faith may help parents in this sample to accept a child affected with a genetic disorder. Providing supportive care to the family and extended family and recognizing the role of religion in this cultural group may reinforce positive coping. When parents do not have extended family, they may be at risk for impaired coping. Also, it may be that having a child who needs technological support (such as wheelchairs or suctioning) engenders different coping strategies. Nonetheless, there seems to be a common theme in most of the previous and this current research that strategies in the family integration subscale are helpful when a child has either a chronic or a genetic diagnosis (Acevedo 2009).

These findings may be the result of the unique sample; Puerto Rican parents in this study were well educated, insured, and received care at a genetics clinic with access to additional specialty services. These findings may also be the result of parental and gender-specific roles within the traditional Hispanic culture: Women manage the home and raise children while men are expected to protect the family and be strong. Role expectations in a Hispanic or Latino community may contribute to higher scores in negative mood, decreased positive coping, or both. Future study is needed to better understand cultural influences on mood and coping in parents and families (Acevedo 2009).

Isn't this fascinating? Once more research shows that what we are and how we react to it has a molecular and clinical component but more so these can be imprinted with the social, ideological, cultural, and religious factors as well.

Prenatal Diagnosis

I would like to put particular interest in the fact that prenatal histories of pregnancies with STD fetuses are remarkably uneventful. More than 80% of cases are not diagnosed prenatally despite prenatal ultrasounds being performed. In our experience the majority of STDs diagnosed before birth are because of a sibling with the condition or significant family history of affected individuals.

The diagnosis of a disorder that can present with significant neonatal complications including respiratory distress and failure and needs to be done as early in pregnancy as possible so that the baby is born in an appropriate hospital facility with a team of medical experts that are knowledgeable about this condition. Failure to do so will result in increased morbidity and mortality. There is no explanation to missing out a prenatal ultrasound of a fetus with STD. Early in the second trimester of pregnancy the multiple segmentation and formation vertebral defects can easily be identified by ultrasound. It is the understanding of this author that an appropriate obstetrical care should include an ultrasound in each trimester of pregnancy.

No prenatal or pre-conceptual history of environmental teratogens or exposures has been associated with these anomalies. No significant first- or second-degree consanguinity has been documented; however, the areas that seem to have the higher incidence of spondylothoracic dysostosis in Puerto Rico are the west and northwest regions of the island. Further ahead in this chapter, we will discuss the molecular aspects of STD in the Puerto Rican population, but it is necessary to mention here that there is evidence of an initial founder effect that accounts for the most frequent mutation, *MESP2* E103X, that produces the classical STD phenotype in homozygous individuals. However, more mutations have been identified in this population, suggesting a potential positive-selection pressure or heterozygote advantage.

The identification of the gene responsible for STD in more than 90% of the cases in Puerto Rican population made possible the provision of prenatal and preconception molecular diagnosis in this population. Prenatal molecular diagnosis is also available to other ethnic backgrounds; however, the percentages of STD due to *MESP2* gene mutation may vary for different ethnic backgrounds.

Molecular Analysis of *MESP2* in Cases of Spondylothoracic Dysostosis

As stated above, segmentation defects of the vertebrae in humans are present in a wide variety of rare but well-characterized disorders, as well as many diverse and poorly understood phenotypic patterns (Turnpenny et al. 2007, Chapter 5). Many of these disorders, including spondylothoracic dysostosis, have been lumped together under the diagnosis of "Jarcho–Levin syndrome," and molecular studies are now being used to produce a new classification scheme (Chapter 5). We will focus on the molecular aspects of spondylothoracic dysostosis (STD) in this chapter.

Spondylothoracic dysostosis has been well characterized as an autosomal recessive disorder with high prevalence in the Puerto Rican population, comprising 49% of the STD cases reported in the medical literature (Cornier et al. 2004). The same phenotype has also been described in other patient populations (Lavy et al. 1966, Mosely and Bonforte 1969, Pochaczevsky et al. 1971, Gellis et al. 1976, Solomon et al. 1978, Tolmie et al. 1987, Schulman et al. 1993, McCall et al. 1994). Patients with STD exhibit a short stature due to multiple defects in vertebral segmentation and spine formation, an increased antero-posterior (AP) thoracic diameter, and, radiologically, a characteristic "crab-like" appearance of the thoracic cage on AP projection, which clearly differentiates it from SCD. The ribs are fused posteriorly at the costovertebral junctions.

Sequencing analysis of the *MESP2* gene demonstrated that mutations in this gene are responsible for the majority of STD cases in the Puerto Rican population. Classical DNA, polymerase chain reaction (PCR), and sequencing analysis were performed in 14 patients with STD and their respective parents (Cornier et al. 2008). A new homozygous nonsense mutation in the *MESP2* gene was identified as the most frequent cause of STD in the Puerto Rican population. This change consisted of a single-base pair substitution mutation (c.307G > T) in the beta helix–loop–helix (bHLH) domain of the *MESP2* gene. The mutation resulted in the replacement of a glutamic acid codon (GAG) at position 103 with a stop codon (TAG) and the creation of a novel *Spe*I restriction enzyme site. The mutation, E103X (p.Glu103X), occurs in exon 1 in the middle of the bHLH domain and is predicted to encode a nonfunctional protein.

An additional patient was heterozygous for the E103X mutation and a novel L125V missense *MESP2* mutation (Cornier et al. 2008). The L125V mutation occurs in a conserved leucine residue in the bHLH domain and is predicted to be deleterious by the Sorting Intolerant From Tolerant (SIFT) prediction program (Ng and Henikoff 2001). Another patient (000408-05) was heterozygous for the E103X mutation, but a second *MESP2* mutation was not identified; neither was there any mutation in *DLL3* or *LFNG* genes (Cornier et al. 2008). It is unclear how the E103X mutation contributes to the STD phenotype in this particular patient. The father of this patient was of Puerto Rican and Chinese ethnicity, which may explain the lack of a second *MESP2* gene mutation. Two variants of the *MESP2* gene, A66G and S174F, were identified in two parental carriers and were believed to be polymorphisms, as these variants were not transmitted to their affected offspring. Further analysis of probands with STD has revealed other mutations such as a novel nonsense mutation, E230X. This mutation (c.688G > T) results in the replacement of glutamic acid at position 230 with a premature termination codon (p.Glu230X). Heterozygous carriers of this mutation were unaffected.

To determine whether these mutations were responsible for the clinical manifestations in STD, transcriptional activities of *MESP2* variants were analyzed using luciferase reporter assays, which allowed us to measure how a mutation may affect the expression of not only the MESP2 protein itself but also other interacting genes (Cornier et al. 2008). Previous studies in mouse revealed that *Mesp2* synergizes with Notch signaling to activate transcription from an *Lfng* reporter in cultured cells (Morimoto et al. 2005). Cornier et al. (2008) confirmed that the human MESP2 also retained similar activity as the mouse *Lfng* reporter, albeit weaker than the mouse *Mesp2*. After examining three different *MESP2* mutant constructs harboring two mutations reported in this chapter, E103X and L125V, as well as the 4-bp duplication (500–503 dup) as reported previously (Whittock et al. 2004), the results indicate that all three mutations lack transcriptional activities in this assay.

We compared the genotype of 10 patients with STD with their respective clinical phenotypes and natural history. The two patients who died during early infancy were homozygous for the E103X mutation, and less clinical measurements and data were available. Axial skeletal dimensions were compared to normal standards (Dimeglio and Bonnel 1990). Phenotypic data suggest that being heterozygous for the E103X mutation produces a milder phenotype in terms of the thoracic measurements although due to the small sample size, the differences were not statistically significant (Cornier et al. 2008).

It has been shown that homozygosity for the *MESP2* E103X mutation causes STD and is a common gene mutation in Puerto Rican patients, highly suggestive of a founder effect (Cornier et al. 2008). Furthermore, two additional mutations – E230X and L125V – which are predicted to alter *MESP2* function have been described. The E230X mutation truncates the protein in the C-terminal domain, and the L125V destroys a conserved leucine residue in the bHLH domain. This suggests that multiple mutations in *MESP2*, including compound heterozygosity, can be associated with STD phenotypes. Heterozygosity for the *MESP2* E103X mutation without, apparently, harboring another *MESP2* mutation has also been described, with no sequence change evident in either *DLL3* or *LFNG*.

It is possible that a partial gene deletion or a mutation in the promoter or regulatory region may cause STD phenotype in patients not harboring mutation in the coding regions of the gene. Alternatively, a patient with STD may carry a mutation in an as-yet unidentified gene. Variations in phenotypes associated with mutations in the same gene might be explained by the position and effect of the mutation in the function of the gene. The E103X and E230X mutations identified in patients with STD are located within the first exon of the *MESP2* gene, and the resulting mutant mRNA transcripts are predicted to be susceptible to nonsense-mediated decay. Therefore, patients who are homozygous or compound heterozygous for these mutations are likely to have a reduced or absent MESP2 protein.

MESP2 and Spondylocostal Dysostosis

A mutation in the *MESP2* gene has also been identified in a small consanguineous family with a milder form of spondylocostal dysostosis (SCD) (Whittock et al. 2004). The change consisted of a 4-bp duplication that traduces clinically in less severe segmentation abnormalities and misaligned ribs. This genetic change occurs after the bHLH domain and causes a frame shift that results in a premature stop codon within the second and final exon of *MESP2*. Transcripts carrying this mutation would not be subject to nonsense-mediated decay. In these patients, a truncated protein containing an intact bHLH domain is predicted to retain some function. This may account for the variable severity of phenotypes associated with different *MESP2* mutations.

The phenotypic data suggest that homozygosity for the E103X mutation gives rise to a more severe phenotype compared with compound heterozygosity involving this mutation (according to thoracic measurements). Results from pulmonary function tests of patients heterozygous for E103X should show a less restrictive pattern than in those who are homozygous. In this study, thoracic CT scans had not been performed on the E103X homozygous patients who had died; thus, comparison with homozygous E103X survivors was not possible.

Role of *MESP2* in Somitogenesis

In the mouse embryo, segmentation is first observed in the PSM as a striped expression of the *Mesp2* gene (Saga et al. 1997). At a defined PSM level, cells respond to a periodic signal from the segmentation clock by activating the genes of the *Mesp2* family in a segment-wide domain (Morimoto et al. 2005). These stripes of *Mesp2* expression provide the blueprint on which the embryonic segments, the somites, will form. Hence, disrupting this process results in severe anomalies of the regular arrangement of vertebrae as is observed in patients with congenital scoliosis. Subsequently, genes of the *Mesp2* family play a critical role in positioning the future somite boundary and in defining the AP polarity of the forming somite (Morimoto et al. 2005). This rostrocaudal subdivision of the somite is critical for the formation of vertebrae, which develop when the caudal half of one somite fuses with the rostral half of the next somite during a process called resegmentation (reviewed in Chapter 2). Our studies suggest that *MESP2* mutations disrupt somitogenesis in humans, resulting in STD and SCD (Figs. 6.8 and 6.9).

Fig. 6.8 AP view of
computer tomography scans
of the whole spine with
three-dimensional
reformation using Xvision
EX Toshiba (Japan)
2 mm cuts

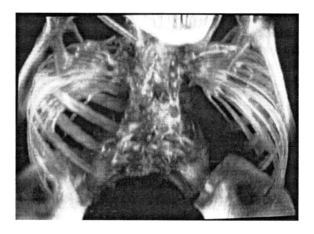

Fig. 6.9 Lateral view of
computer tomography scans
of the whole spine with
three-dimensional
reformation using Xvision
EX Toshiba (Japan) 2 mm
cuts. Bone fusions may
extend to the lateral aspects
of the thorax

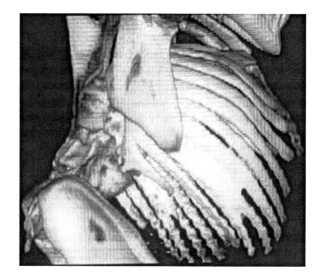

Clinical Management of Patients with STD

Given the high incidence of a disorder that is otherwise rare elsewhere has provided
us the opportunity to study the natural history of spondylothoracic dysostosis and
to design medical strategies that improve the outcome of patients with this disorder.
We will divide this into three different stages according to age.

Neonatal Period

- It is of extreme importance to identify these babies prenatally so that an expert health-care team is ready for their birth. This team should include an obstetrician, anesthesiologist, neonatologist, geneticist, nurses, dietician, genetic counselor, and pediatrician.
- If the baby presents with respiratory difficulties, surfactant factor should be provided immediately as per the guidelines. Not all babies with STD will have respiratory difficulties at birth; approximately 65% of them do have some type of respiratory problem in the newborn period ranging from mild distress requiring indirect oxygen supplementation to frank uncompensated respiratory acidosis requiring mechanical ventilation.
- Elevated CO_2 levels may present in the newborn period (hypercapnea). Babies with STD seem to tolerate hypercapnea relatively well. Patients with CO_2 levels as high as 60–70 have improved their respiratory status with CPAP (constant positive air pressure) avoiding intubations. Follow-up of these patients for over 10 years has shown no neurological or developmental impairment.
- If mechanical ventilation is unavoidable, ventilator pressure should remain as low as possible to avoid hypoxia and correct respiratory acidosis. The ventilatory problem of these patients is not due to alveoli gas exchange but rather to the incapacity of mechanical expansion of the thorax and secondary lung hypoplasia in some cases. Mechanical ventilation should be discontinued as soon as possible.
- Constant cardio-respiratory monitoring should always be pursued.
- Avoid bolus feeding. Bolus feeding will expand gastric cavity in these babies and so the diaphragm will have more difficulty to be used. Respiration of patients with STD depends solely on the work of the diaphragm muscle since the accessory intercostal muscles are impeded from work due to the extensive rib fusion of the thoracic cage. Slow nasogastric feedings should be considered instead at a rate of approximately 0.75–1 ounce/h for 16–20 h with 20 calories per ounce baby formula and a rest period of 4–6 h. This will provide 125–147 kcal/day to babies that weigh between 2.72 and 3.18 kg at birth, which should be enough to meet the babies' needs in the newborn period. More concentrated formulas may be used to optimize caloric intake.
- Be aggressive in the treatment of any infectious agent.

Infancy and Early Childhood

- Be aggressive in the treatment of upper respiratory tract ailments. This includes antibiotic therapy as needed.
- Babies with STD should be preventively vaccinated against the RSV (respiratory syncytial virus). This immunization should be performed year long (not only in the RSV season) in the first 3–5 years of life.

- Respiratory therapy with bronchodilators may be considered in the management of upper respiratory tract infections with or without bronchoconstriction (wheezing).
- Avoid mechanical ventilation unless strictly necessary. Patients in infancy and childhood do seem to tolerate well-elevated levels of CO_2 without significant clinical complications. Close and constant cardio-respiratory monitoring should always be pursued.
- Physical and occupational therapies may be useful to provide support to the development of these patients.

Late Childhood and Adulthood

- Be aggressive in the medical management of respiratory tract infections.
- Orthopedic follow-up should include the presence of scoliosis, spine malrotations, and bone bars. These may need to be addressed surgically if the malformation is asymmetrical.
- Watch for osteoporosis. There are preliminary data that suggest that patients with STD develop osteoporosis in their early forties. It is not clear if this finding is secondary to any endocrine abnormality or to lack of strenuous physical activity throughout life.

Acknowledgments I would like to acknowledge the dozens of patients with STD and their families for all their help and collaboration in establishing the natural history of this disease as well as determining the molecular etiology. They have become part of my extended family and I feel privileged for their "cariño y respeto." "Yo también los quiero mucho."

I would also like to acknowledge ICVAS, the International Consortium of Vertebral Abnormalities and Scoliosis, as well as my collaborators Dr. Norman Ramirez, Dr. Jose Acevedo, Dr. Simon Carlo, Dr. Jesse Romeu, Dr. Jhon Flynn, Dr. Sandra Arroyo, Dr. Hostos Fernandez, Lcda. Ivette Ramos, and Lcda. Velma Franceschini for their support and collaboration in the research of STD throughout the years.

Finally, I would like to express my depth of appreciation to the editors of this fantastic book, Drs. Kenro Kusumi and Sally Dunwoodie, for inviting me to share with you the history, research, and personal experience on spondylothoracic dysostosis syndrome.

To all of you **"gracias mil."**

References

Acevedo, J.R. 2009. Coping Patterns and Mood States in Puerto Rican Parents of Children with a Genetic Disorder. Unpublished Thesis. Case Western Reserve University, Cleveland, Ohio.
Ayme, S. and Preus, M. 1986 Spondylocostal/spondylothoracic dysostosis: the clinical basis for prognosticating and genetic counseling. Am. J. Med. Genet. 24:599–606.
Cantu, J.M., Urrusti, J., Rosales, G., and Rojas, A. 1971. Evidence for autosomal recessive inheritance of costovertebral dysplasia. Clin. Genet. 2:149–154.
Casamassima, A.C., Casson-Morton, C., Nance, W.E., Kodroff, M., Caldwell, R., Kelly, T., and Wolf, B. 1981. Spondylocostal dysostosis associated with anal and urogential anomalies in a Mennonite sibship. Am. J. Med. Genet. 8:117–127.

Cornier, A.S., Ramirez, N., Arroyo, S., Acevedo, J., Garcia, L., Carlo, S., and Korf, B. 2004. Phenotype characterization and natural history of spondylothoracic dysplasia syndrome: a series of 27 new cases. Am. J. Med. Genet. 128:120–126.

Cornier, A.S., Staehling-Hamptom, K., Delventhal, K.M., Saga, Y.,Caubet, J.F., Sasaki, N., Ellard, S., Young, E., Ramirez, N., Carlo, S., Torres, J., Emans, J.B., Turnpenny, P.D., and Pourquié, O. 2008. Mutations in the MESP2 gene cause spondylothoracic dysostosis/Jarcho-Levin syndrome. Am. J. Hum. Genet. 82(6):1334–1341.

Dimeglio, A. 1993. Growth of the spine before 5 years. J. Pediatr. Orthop. B 1:102–107.

Dimeglio A. 2005. Growth in pediatric orthopaedics. In Lovell and Winter's Pediatric Orthopaedics, 6th Ed. eds. T. Morrissy, and Weinstein, S.L., pp. 35–65. Philadelphia, PA: Lippincott Williams and Wilkins.

Dimeglio. A., Bonnel. F., Le rachis en croissance. Paris: Springer; 1990.

Floor, E., De Jong, R.O., Fryns, J.P., Smulders, C., and Vles, J.S.H. 1989. Spondylocostal dysostosis: an example of autosomal dominant transmission in a large family. Clin. Genet. 36:236–241.

Franceschini, P., Grassi, E., Fabris, C., Bogetti, G., and Randaccio, M. 1974. The autosomal recessive form of spondylocostal dysostosis. Pediatr. Rad. 112:673–676.

Gellis, S.S., Feingold, M., and Pashayan, H.M. 1976. Picture of the month: the EEC syndrome. Am. J. Dis. Child. 130:653–654.

Herold, H.Z., Edlitz, M., and Baruchin, A. 1988. Spondylothoracic dysplasia: a report of ten cases with follow-up. Spine 13:478–481.

Jarcho, S. and Levin, P. 1938. Hereditary malformation of the vertebral bodies. Bull. Johns Hopkins Hosp. 62:216–226.

Lavy, N.W., Palmer, C.G., and Merritt, A.D. 1966. A syndrome of bizarre vertebral anomalies. J. Pediatr. 69:1121–1125.

McCall, C.P., Hudgins, L., Cloutier, M., Greenstein, R.M., and Cassidy, S.B. 1994. Jarcho-Levin syndrome: unusual survival in a classical case. Am. J. Med. Genet. 49:328–332.

Martínez-Cruzado, J.C., Toro-Labrador, G., Viera-Vera, J., Rivera-Vega, M.Y., Startek, J., Latorre-Esteves, M., Román-Colón, A., Rivera-Torres, R., Navarro-Millán, I.Y., Gómez-Sánchez, E., Caro-González, H.Y., and Valencia-Rivera, P. 2005. Reconstructing the population history of Puerto Rico by means of mtDNA phylogeographic analysis. Am. J. Phys. Anthropol. 128(1):131–155.

Morimoto, M., Takahashi, Y., Endo, M., and Saga, Y. 2005. The Mesp2 transcription factor establishes segmental borders by suppressing Notch activity. Nature 435:354–359.

Moseley, J.E. and Bonforte, R.J. 1969. Spondylothoracic dysplasia: a syndrome of congenital anomalies. Am. J. Roentgen. 106:166–169.

Ng, P.C. and Henikoff, S. 2001. Predicting deleterious amino acid substitutions. Genome Res. 11:863–874.

Pochaczevsky, R., Ratner, H., Perles, D., Kassner, G., and Naysan, P. 1971. Spondylothoracic dysplasia. Radiology 98:53–58.

Ramirez, N., Flynn, J.M., Serrano, J.A., Carlo, S., and Cornier, A.S. 2009. The Vertical Expandable Prosthetic Titanium Rib in the treatment of spinal deformity due to progressive early onset scoliosis. J. Pediatr. Orthop. B 18(4):197–203.

Rimoin, D.L., Fletcher, B.D., and McKusick, V.A. 1968. Spondylocostal dysplasia: a dominantly inherited form of short trunk dwarfism. Am. J. Med. Genet. 45:948–953.

Saga, Y., Hata, N., Koseki, H., and Taketo, M.M. 1997. Mesp2: a novel mouse gene expressed in the presegmented mesoderm and essential for segmentation initiation. Genes Dev. 11: 1827–1839.

Schulman, M., Gonzalez, M.T., and Bye, M.R. 1993. Airway abnormalities in Jarcho-Levin syndrome: a report of two cases. J. Med. Genet. 30:875–876.

Solomon, L., Bosh-Jimenes, R., and Reiner, L. 1978. Spondylothoracic dysostosis. Arch. Pathol. Lab. Med. 102:201–205

Tolmie, J.L., Whittle, M.J., McNay, M.B., Gibson, A.A., and Connor, J.M. 1987. Second trimester prenatal diagnosis of the Jarcho-Levin syndrome. Prenat. Diagn. 7:129–134.

Turnpenny, P.D., Alman, B., Cornier, A.S., Giampietro, P.F., Offiah, A., Tassy, O., Pourquie, O., Kusumi, K., and Dunwoodie, S. 2007. Abnormal vertebral segmentation and the notch signaling pathway in man. Dev. Dyn. 236:1456–1474.

Whittock, N.V., Sparrow, D.B., Wouters, M.A., Sillence, D., Ellard, S., Dunwoodie, S.L., and Turnpenny, P.D. 2004. Mutated MESP2 causes spondylocostal dysostosis in humans. Am. J. Hum. Genet. 74:1249–1254.

Chapter 7
Progress in Understanding Genetic Contributions in Syndromic and Non-Syndromic Disorders Associated with Congenital, Neuromuscular, and Idiopathic Scoliosis

Philip F. Giampietro

Introduction

Vertebral development occurs through a sequential and highly orchestrated series of interconnected events involving fibroblast growth factor (FGF), WNT, Notch, and transforming growth factor beta (TGF-β) receptor signaling pathways. Perturbations in these pathways can result in the development of both congenital (curvature of the spine due to an abnormality in vertebral formation) and idiopathic scoliosis (spinal curvature associated with normal vertebral morphology and in the absence of secondary causes such as Marfan syndrome, chromosome abnormality, or neuromuscular etiology). This chapter will focus on syndromic conditions that are associated with scoliosis and how understanding these conditions may contribute to identification of genes for idiopathic scoliosis. A large number of syndromes are associated with idiopathic scoliosis, resulting in connective tissue alteration or a neuromuscular basis for spinal curvature. Although various candidate loci have been recognized, currently very few susceptibility genes for the development of idiopathic scoliosis have been identified. It is probable that abnormalities in the development and postnatal maintenance of the extracellular matrix may contribute to pathogenesis of idiopathic scoliosis. Marfan and Stickler syndromes are both associated with scoliosis and are caused by mutations in *FBN1, COL2A1, COL11A1, COL11A2,* and *COL9A1*. Despite the abundance of these genes in connective tissue, there is currently no compelling evidence for association of these genes with idiopathic scoliosis. The presence of both vertebral malformations and scoliosis without vertebral malformations in Smith–Magenis syndrome suggests that the *RAI1* (retinoic acid-induced 1) gene may play a role in the development of both congenital and idiopathic scoliosis. Additional relevant syndromes and animal models which represent the potential etiological mechanisms operative in scoliosis will

P.F. Giampietro (✉)
Department of Pediatrics, University of Wisconsin, Madison, WI, USA; Department of Medical Genetic Services, Marshfield Clinic, Marshfield, WI USA
e-mail: pfgiampietro@pediatrics.wisc.edu

K. Kusumi, S.L. Dunwoodie (eds.), *The Genetics and Development of Scoliosis,* DOI 10.1007/978-1-4419-1406-4_7, © Springer Science+Business Media, LLC 2010

be reviewed. Analysis of the current understanding of these conditions and models will be employed to promote insight into the genetically mediated, biochemical and neurological susceptibilities contributing to idiopathic scoliosis. Approaches to understand the etiology or etiologies for congenital scoliosis which is not associated with an underlying syndrome will also be discussed.

Epidemiology of Segmentation Defects of Vertebrae (SDV)

Spine development is an intricate process orchestrated by complex interactions of multiple signaling genes of the Wnt, Fgf, TGF-β, and Notch signaling pathways. This chapter will focus on the genetic aspects related to the incidence of sporadic and familial SDV syndromes. Congenital vertebral malformations encompass defects that might arise in segmentation, leading to segmentation defects of the vertebrae (SDV), as well as later embryological disruptions in sclerotomal resegmentation and fusion. Recent developments in the understanding of the genetic etiology of syndromic and nonsyndromic vertebral malformation syndromes will be discussed. In addition, potential genetic mechanisms underlying the etiology of idiopathic scoliosis and the genetic inter-relationships between congenital and idiopathic scoliosis will be examined.

SDV in humans represents a significant health problem that may cause kyphosis and/or scoliosis, resulting in back and neck pain, disability, cosmetic disfigurement, and functional distress. Although estimates indicate a prevalence of approximately 0.5–1/1000, the true incidence of vertebral malformations is unknown (Shands and Eisberg 1955, Wynne-Davies 1975, McMaster and Ohtsuka 1982, McMaster and Singh 1999).

Sporadically Occurring SDV Syndromes

Vertebral malformations may represent an isolated finding, occur in association with other renal, cardiac, or spinal cord malformations, or occur as part of an underlying chromosome abnormality or syndrome at an estimated frequency of 30–60% (Jaskwhich et al. 2000). Several commonly encountered or known syndromes associated with SDV are described in detail below.

Some commonly occurring syndromes are as follows.

Alagille Syndrome

Alagille syndrome includes facial dysmorphology characterized by a broad forehead and deeply set eyes, bile duct abnormalities, pulmonic stenosis, and vertebral malformations (Alagille et al. 1975). The principal features of Alagille syndrome include bile duct paucity and abnormalities of the heart, eye, kidney, pancreas, skeleton, and facial dysmorphism (Alagille et al. 1975). Pulmonic stenosis is the most common cardiac abnormality (Emerick et al. 1999). Facial features may include a broad forehead with deeply set eyes, moderate hypertelorism, a straight or saddle-shaped nose with a broadened nasal tip and pointed chin (Kamath et al. 2002). Ophthalmologic abnormalities commonly include anterior chamber defects such as Axenfeld anomaly, Reiger anomaly, and retinal pigment changes and the persistence of posterior embryotoxon (Hingorani et al. 1999). Butterfly vertebral anomalies appearing as vertebral clefts are observed in 22.8% of affected individuals and represent errors in somitogenesis.

Inheritance is autosomal dominant due to haploinsufficiency (Alagille et al. 1975). *JAG1* mutations have been identified in approximately 70% of patients with Alagille syndrome(Colliton et al. 2001). Mutations in *NOTCH2*, a second gene, have been observed in patients with Alagille syndrome with severe renal manifestations (McDaniell et al. 2006). These mutations include small and entire gene deletions, nonsense, missense, truncating, and splicing mutations. As a ligand of the Notch receptor, *JAG1* plays a central role in developmental regulation. Notch undergoes proteolysis following ligand binding and releases a C-terminal fragment that translocates to the nucleus, resulting in target gene transcription. The observation that double heterozygous mice for *Jag1* null allele and a *Notch2* hypomorphic allele show phenotypic abnormalities consistent with Alagille syndrome supports interaction between Jagged1 and Notch2 (Xue et al. 1999).

Klippel–Feil Anomaly

Klippel–Feil is defined by the presence of cervical vertebral fusion abnormalities (Samartzis et al. 2006). First reported in 1894 (Hutchinson 1894), followed by Klippel and Feil's description of a French tailor with massive cervical fusion who died of renal disease (Klippel and Feil 1912), Klippel–Feil anomaly, sometimes classified as a syndrome, is a disorder of the cervical spine characterized by faulty segmentation of vertebrae in the cervical region of the spinal column. While the majority of cases represent sporadic occurrences within a particular family, autosomal dominant, autosomal recessive, and X-linked forms of Klippel–Feil anomaly have been reported (Klippel and Feil 1912, Feil 1919, Hensinger and MacEwen 1982, Clarke et al. 1998, Manaligod et al. 1999, Samartzis et al. 2006). Additional multi-organ system malformations, including neural tube defects, thoracic cage abnormalities, pulmonary, cardiovascular, other skeletal anomalies, genitourinary abnormalities, myopathy, neuropathy, and cognitive disorders, may occur in association with Klippel–Feil anomaly (reviewed in Tracy et al. 2004).

A pericentric inversion inv(8)(q22.2q22.3) segregating in a four generation family with affected members with Klippel–Feil anomaly (Clarke et al. 1995) has led to the identification of mutations in *GDF6*, a member of the BMP family, in familial and sporadic cases of Klippel–Feil anomaly. In a three-generation family with autosomal dominant Klippel–Feil anomaly, a missense mutation in c.746C > A (p.A249E) was identified (Tassabehji et al. 2008). In addition, a recurrent mutation in a highly conserved residue c.866T > C (p.L289P) of GDF6 was identified in 2 of 121 sporadic cases of Klippel–Feil anomaly (Tassabehji et al. 2008). The sporadic cases included a female fetus with multiple vertebral segmentation abnormalities encompassing the entire spine, Arnold Chiari malformation, and rocker bottom feet. Genetic heterogeneity for Klippel–Feil anomaly is supported by the occurrence of *GDF6* mutations in only 2 of 121 patients studied. The understanding of disease transmission and clinical impact is enhanced by the availability of cases representing both familial and sporadic occurrences.

There is evidence that mutations in the *PAX1* gene contributes to the occurrence of Klippel–Feil anomaly (McGaughran et al. 2003). *Pax1* is expressed in the sclerotome, which gives rise to the vertebrae and ribs. The identification of *Pax1* mutations in the mouse *undulated* provides evidence that sclerotome condensation is a *Pax1*-dependent process (Dietrich and Gruss 1995). Medial sclerotome condensation fails to occur at the lumbosacral level in the mouse mutant *undulated,* thus preventing formation of intervertebral discs and vertebral bodies. Mutations in *PAX1* were identified in 3 of 63 patients diagnosed with Klippel–Feil anomaly. One of these patients, and the asymptomatic mother of the patient revealed a (CCC > GCC; P61A) missense mutation 38 bp upstream from the paired-box region, which is associated with DNA binding of the PAX1 protein. Assuming this mode of inheritance is autosomal dominant, reduced penetrance could be postulated if this mutation promotes causation.

Hemifacial Microsomia (Oculo-Auriculo-Vertebral Spectrum)

The spectrum of oculo-auriculo-vertebral spectrum (OAVS) includes craniofacial abnormalities, ear, ocular, cardiac, skeletal, and renal malformations (Rollnick et al. 1987). In most instances, hemifacial microsomia and Goldenhar syndrome represent sporadic occurrences within a particular family, although autosomal dominant and recessive inheritance has been reported (Burck 1983, Stoll et al. 1998). Presently, no locus for OAVS has been identified and a vascular disruption hypothesis has been proposed (Poswillo 1973). Evidence for an epigenetic cause for OAVS has been identified in a patient with 46,XX,t(4:8)(p15.3;q24.1) with the breakpoint on chromosome 4 close to the *BAPX1* gene (Fischer et al. 2006). *BAPX1*, a member of the NK2 class of homeobox genes, is an early marker of chondrogenesis and is expressed initially in the posterior foregut at E8.0 in the mouse embryo (Tribioli et al. 1997, Yoshiura and Murray 1997). *Bapx1* expression is restricted to the sclerotomal region of the somite which gives rise to the axial skeleton. This is followed by a further expression domain at stage E9.5 within the inferior portion of the first branchial arch ultimately expanding into developing structures of the craniofacial skeleton by E10.5. At E16.5 *Bapx1* is expressed in the axial skeleton and in a discrete domain within the mandibular component (Meckel's cartilage) of the first branchial arch. Although no mutation in *BAPX1* was identified following DNA sequence analysis in 105 patients, strong allelic expression imbalance (Fischer 2006) was observed in fibroblast cultures of 5 of 12 patients and was absent in

9 control fibroblast cultures. Treatment with the histone deacetylase inhibitor tri-chostatin A resulted in reactivation of the downregulated allele, suggesting that histone acetylation is important for development of the first and second branchial arch derivatives. Additionally, the sAEI allele was found only in parents and their relatives, suggesting that the sAEI allele predisposes to OAVS.

VACTERL Syndrome

VACTERL (defined by the presence of three or more of the following, Vertebral, Anal atresia, Cardiac, TracheoEsophageal fistula, Renal, Limb anomalies) syndrome (Botto et al. 1997) is a condition which usually represents a sporadic occurrence within a given family.

Genetic etiologies have been considered for the VACTERL syndrome, which almost always occurs sporadically. A patient with omphalocele and VACTERL-related multiple congenital anomalies, including tracheoesophageal atresia, duo-denal atresia, patent ductus arteriosus, right renal agenesis, and limb anomalies of the right hand and fingers, was found to harbor a 3-nucleotide deletion in the 3'UTR region of the PITX2 gene, a member of the paired-type bicoid-related homeobox-containing proteins (Katz et al. 2004). Evidence for respiratory chain defects has been identified in two patients with VACTERL syndrome. The first patient exhibited esophageal atresia, vertebral and rib anomalies, a radial ray defect, liver dysfunction characterized by elevated lactate, and episodes of hypoglycemia and was ultimately found to have complex IV respiratory chain deficiency (Thauvin-Robinet et al. 2006). A 3243G point mutation was also identified in a second patient with VACTERL syndrome. These observations provide evidence for contributions of both patterning and antenatal respiratory chain defects to the development of VACTERL syndrome, both consistent with sporadic occurrence.

A de novo 21-base pair heterozygous deletion (c.163_183del) resulting in removal of seven alanines (A55_A61del) from the polyalanine tract of the HOXD13 gene was identified in a 17-year-old female with clinical features of VACTERL syndrome including imperforate anus, tetralogy of Fallot, hydronephro-sis, and hydroureters resulting from bilateral vesicoureteric reflux (Garcia-Barcelo et al. 2008). The patient also had fusion of the distal interphalangeal joints of the fourth and fifth toes. Since HOXD13 is a sonic hedgehog gene target during gastrointestinal development and mutations in HOXD13 are associated with limb abnormalities, HOXD13 is strong candidate gene for VACTERL syndrome.

Currarino Syndrome

The triad of sacral formation defect, hindgut malformation, and a presacral mass consisting of an anterior meningocele, teratoma, rectal duplication, or a combination of these is termed "Currarino syndrome" (Currarino et al. 1981). The first sacral vertebrae is intact ("sickle-cell sacrum") with sacral agenesis below S2. Currarino syndrome is caused by mutations in the *HLXB9* gene (Ross et al. 1998, Belloni et al. 2000). This gene encodes a transcription factor, HB9 protein, which belongs to the Mnx-class homebox genes (Ross et al. 1998) encoding proneuronal basic helix–loop–helix (bHLH) proteins such as *Ngn2*, NeuroM, and LIM domain proteins such as *Isl1*, *Lhx3*, and *Lhx4* (Lee and Pfaff 2003, Lee et al. 2004, Nakano et al. 2005). It is noteworthy that 2 of 26 patients with *HLXB9* mutations identified in one series did not display sacral agenesis (Crétolle et al., 2008).

Because of the complexity of the development of the caudal end of the embryo, sacral agenesis is rarely isolated (Pang 1993). Other syndromic associations of sacral agenesis include the OEIS complex (*O*mphalocele, cloacal *E*xstrophy, *I*mperforate anus, and *S*pinal deformities), VACTERL syndrome (Harlow et al. 1995). Partial or total absence of the sacrum is associated with diabetic embryopathy (Blumel et al. 1959). In a series of 48 cases with caudal regression syndrome, no mutations in *HLXB9* were identified, suggesting that other genes and/or environmental factors are involved in its pathogenesis (Merello et al. 2006).

Teratogens: A Significant Etiologic Factor in the Development of SDV

The influence of various teratogens on the expression of genes associated with vertebral development has not been well studied. One possible model for the development of SDV is multifactorial and may reflect underlying genetic alterations in maternal and/or fetal genes associated with vertebral development. An understanding of the possible environmental impact on vertebral development may help to provide more complete genetic counseling for families.

The occurrence of SDV in humans has been reported in association with alcohol (Tredwell et al. 1982), anticonvulsant medications including valproic acid (Band et al. 2002) and dilantin (Hanold 1986), hyperthermia (Edwards 1986), and maternal insulin-dependent diabetes mellitus (Ewart-Toland et al. 2000). The association of spinal cord anomalies in humans and neural tube defects in mice with congenital scoliosis, which will be described in this chapter, raises the possibility that similar etiological mechanisms underlie development of these malformations in both species. Folic acid use has been associated with the prevention of neural tube defects (Lewis et al. 1998). Further studies are required to determine the impact of antenatal folic acid use on development of SDV.

An association of vertebral malformations was reported in Sprague-Dawley rat fetuses exposed to I(Kr)-blockers (class III antiarrhythmic agent) *in utero* (Sköld et al. 2001). The proposed teratogenic mechanism postulates that vertebral malformation is mediated by a temporary induction of hypoxia and reoxygenation injury via the induction of embryonic cardiac arrhythmia. Alteration of patterning gene expression in homeobox genes and sonic hedgehog has been observed in conjunction with maternal exposures to carbon monoxide (Farley et al. 2001) and boric acid (Wéry et al. 2003).

Fumonisins, environmental toxins produced by molds *Fusarium moniliforme* (*F. verticillioides*) *F. proliferatum* and other *Fusarium* species, have been associated with decreased ossification of vertebral bodies in newborn rats whose mothers were injected with purified fumonisins B1 (Lebepe-Mazur et al. 1995). Severe vertebral anomalies have been observed in pregnant rats who were fed a zinc-deficient diet from days 3–15 of gestation (Hickory et al. 1979). Absence of thoracic and caudal vertebrae in mice has been associated with exposure to chlorpyrifos, an organophosphate pesticide, during pregnancy (Tian et al. 2005). Vertebral deformities and abnormal mechanical vertebral properties have been associated with exposure of juvenile fourhorn sculpin, *Myoxocephalus quadricornis* L. to tetrachloro-1,2-benzoquinone, a component in bleached kraft mill effluents (Bengtsson et al. 1988).

Overview of the Types of Inheritance Patterns Associated with SDVs and Methodologies Used to Identify Mutations in Vertebral Malformation Syndrome Genes

Autosomal recessive inheritance patterns have been reported in spondylocostal dysostosis (described further in Chapter 5) and spondylothoracic dysostosis (STD). Most isolated SDVs and SDVs associated with syndromes represent a sporadic occurrence within a particular family and little information has been obtained regarding their underlying genetic mechanism. Traditional patterns of Mendelian inheritance have been noted in some families with SDV. A large family was identified with many members affected with Klippel–Feil syndrome-associated anomalies displaying an autosomal dominant pattern of inheritance (Clarke et al. 1995). Cytogenetic analysis demonstrated the presence of a paracentric inversion inv(8)(q22.2q23.3) segregating in those individuals with phenotypic manifestations of Klippel–Feil syndrome (Tassabehji et al. 2008). This information was subsequently used to identify mutations in *GDF6* in both sporadic and familial cases of Klippel–Feil syndrome. Additional cytogenetic evidence for loci contributing to the development of SDVs includes the observation of additional chromosomal rearrangements, including de novo balanced reciprocal translocation – t(5;17)(q11.2;q23) (Fukushima et al. 1995), de novo pericentric inversion inv(2)(p12q34) (Papagrigorakis et al. 2003), and translocation – t(5;8)(q35.1;p21.1) (Goto et al. 2006). Traditional linkage methodologies utilizing single nucleotide polymorphisms (SNP) or short tandem repeat polymorphisms (STRPs) may be utilized if large families with SDVs are encountered and cytogenetic etiologies have been eliminated.

Spine development occurs through a harmonic convergence of multiple pathways and several dozen genes. It is likely that mutations in genes in the FGF, WNT, and NOTCH signaling pathways are present in many of the genes in these pathways and have been detected through DNA sequencing methodologies. Phenotypic and genetic heterogeneity represent challenges in elucidating the sequence variation which potentially contributes to the pathogenesis of SDVs. The utilization of a candidate gene approach to identify mutations contributing to SDVs assumes that SDVs have a similar genetic etiology in humans as compared to other species. However, overlapping functions among genes in pathways which potentially contribute to vertebral malformation are likely and would result in several possible genetic etiologies for a particular vertebral malformation. It is also difficult to identify large cohorts of patients to study who have the same or similar types of SDVs in order to rigorously examine association of a mutation in a particular gene with a specific type of SDV.

Array-based comparative genomic hybridization (CGH) is a technique that was developed in order to determine alterations in dosage distribution of small DNA segments throughout the entire genome (Pinkel et al. 1998). There are many human diseases which are associated with genes in which there are copy number changes including Charcot–Marie–Tooth type 1A (*PMP22*), Smith–Magenis syndrome (*RAI1, PLP1*), Pelizaeus–Merzbacher syndrome (*PLP1*), spinal muscular atrophy (*SMN1*), and Rett syndrome (*MECP2*) (Cohen 2007). Examples of such conditions which were rarely detected prior to the advent of CGH arrays include Albright osteodystrophy-like syndrome due to del2q37 (Phelan et al. 1995) and Phelan-McDermid syndrome due to microdeletion 22q11.3 (Phelan 2008). Array-based CGH has also successfully identified a major gene, *CHD7*, which is responsible for CHARGE syndrome, a multiple congenital anomaly (MCA) syndrome that is defined by *C*oloboma of the eye, *H*eart defects, *A*tresia of the choanae, *R*etardation of growth and/or development, *G*enital and/or urinary abnormalities, and *E*ar abnormalities (Vissers et al. 2004). Array-based CGH has the advantage of being able to identify regions of microaneuploidy associated with vertebral malformations across the entire genome, instead of limited exploration of only one focused region. Because etiology of SDV is heterogeneous as documented by preliminary CGH data presented herein and may involve multiple genetic defects, many of which remain to be identified, this approach represents an excellent screening tool for defining additional loci that may harbor genetic defects that underlie etiology of these disorders, although it may lack power to identify a single candidate gene.

Traditional linkage analysis is not a viable approach to identify causative genes since isolated SDV most often represents a sporadic occurrence within a particular family. Candidate gene analyses offer a reasonable alternative method of study. Using mouse–human synteny analysis, 27 eligible loci, 21 of which cause vertebral malformations in the mouse, have been identified (Giampietro et al. 1998, 2003). In a phenotypically well-defined cohort of patients with SDV six candidate genes extrapolated from murine models have been chosen for analysis (Ghebranious et al. 2006, Ghebranious et al. 2007, Ghebranious et al. 2008, Giampietro et al. 2005, Giampietro et al. 2006). The vertebral malformations represented among this cohort spanned the entire length of the spine and are described in greater detail in a prior communication (Giampietro et al. 2005).

These genes were explored in a cohort of patients with isolated, sporadically occurring vertebral malformations: (1) *PAX1*, a patterning gene involved in sclerotome formation and differentiation, (2) *WNT3A*, a major gene essential to segmentation which regulates WNT signaling (Ghebranious et al., 2007), (3) *DLL3*, is a ligand of Notch which inhibits signaling (Ladi et al. J Cell Biol 2003), (4) *SLC35A3*, which underlies complex vertebral malformation syndrome in cattle (Thomsen et al., 2006), (5) *T(brachyury)* (also known as *T*), a transcription factor essential to mesodermal development (Kispert and Herrmann 1994), and (6) TBX6 (Watabe-Rudolph et al. 2002). Potentially pathogenic DNA sequence variants and corresponding phenotypic information are summarized in Table 7.1.

The c.1013C > T variant of the *T* gene, which encodes for the A338V amino acid substitution, displays the most compelling evidence of pathogenicity because of its statistically significant association with SDV ($P = 9.87 \times 10^{-4}$), its presence in three unrelated individuals, its absence in a large reference population and evolutionary conservation in a large reference population (Ghebranious et al. 2008). This variant lies within the second transactivation domain (Ghebranious et al. 2008). The vertebral malformation phenotypes differed among the patients and in each family, and notably the phenotypically unaffected parent also harbored the mutant allele (Ghebranious et al. 2008). The c.1013C > T variant has also been reported in a patient with sacral agenesis and a clinically unaffected parent, providing further evidence for a function of *T* in vertebral morphogenesis (Papapetrou et al. 1999).

Consistent with the expression of *T* throughout the notochord, *T* mutations in mice affect not only the tail, but also more anterior vertebral segments (Searle 1966, Park et al. 1989), spinal cord (Park et al. 1989, Bogani et al. 2004), genitourinary system (Park et al. 1989, Lyon 1996), and heart (Chesley 1935, Abe et al. 2000). While some investigators have reported an association between a C to T transition in intron 7 of *T* and increased risk of neural tube defects (Morrison et al. 1996, Jensen et al. 2004), this finding has not been uniformly reproduced (Shields et al. 2000,

Table 7.1 Potentially pathogenic DNA sequence variants associated with SDV. Reprinted with permission: Giampietro et al. (2008). Molecular diagnosis of vertebral segmentation disorders in humans. Expert Opin. Med. Diagn. 2, 1107–1121

Gene	Phenotypic features	DNA	Amino acid	Frequency
PAX1	T 11 wedge	c.1229C > T	Pro410Leu	1/170
	Multiple thoracic and lumbar SDVs, VSD, polydactyly	c.1238C > T	Pro413Leu	6/2312
DLL3	T5–T6 block, VACTERL	c.805G > A	Gly269Arg	0/174
WNT3A	Supernumerary hemivertebra between T12 and L1	c.400G > A	Ala134Thr	3/890
SLC35A3		None		
TBX6		None		
T	Sacral agenesis	c.1013C > T	A338V	0/886
T	Klippel–Feil			
T	Multiple cervical and thoracic vertebral malformations			

SDV, segmental defects of vertebrae; VSD, ventricular septal defect.

Richter et al. 2002, Speer et al. 2002). Mutant *T* phenotypes should be broadened to include neural tube defects, as well as SDV. The lack of phenotypic expression in the parents harboring the c.1013C > T mutation and the observation that most SDVs represent a sporadic occurrence could be explained by a "multi-hit" kinetic hypothesis for SDV (Knudson 1971). The second event could be environmentally mediated or represent an epigenetic alteration. This hypothesis is supported by the observation that homozygous mutations in *Sp5*, a novel mouse gene that codes for a Cys2His2 zinc finger DNA-binding protein of the Sp1 class, result in increased phenotypic severity of *T/–*, whereas no Sp5-targeted homozygous mutant mice show no skeletal abnormalities (Harrison et al. 2000). Decreased *T* function would enable phenotypic expression of mutations in other genes active during somitogenesis, which would normally be buffered by normal T protein.

Many genetic syndromes associated with idiopathic scoliosis result in connective tissue alteration or a neuromuscular basis for spinal curvature. Although various loci have been identified for idiopathic scoliosis, very few susceptibility genes for the development of idiopathic scoliosis are presently known. It is probable that abnormalities in the development and postnatal maintenance of the extracellular matrix underlie development of idiopathic scoliosis. Below we examine several representative syndromes associated with scoliosis which have a connective tissue, neuromuscular, or mixed etiology. These syndromes may provide clues to its pathogenesis.

Marfan syndrome, Ehlers–Danlos syndrome, spondylocarpotarsal syndrome, and Stickler syndrome are disorders of connective tissue which are associated with scoliosis. Despite the identification of molecular defects in these conditions, mutations in these genes have not been identified as playing a significant role in understanding the pathogenesis of idiopathic scoliosis.

However, these conditions remain important to discuss since their elucidation has led to a better understanding of the extracellular matrix, and can lead to applications essential for the understanding of idiopathic scoliosis. The potential role in for fibronectin in congenital scoliosis will also be discussed.

Marfan Syndrome

Marfan syndrome is an autosomal dominant pleiotropic disorder of connective tissue caused by fibrillin-1 mutations (Dietz et al. 1991). Aortic root dilatation, which can lead to aortic dissection, ectopia lentis, and dural ectasia, represent the major clinical findings observed in Marfan syndrome. Marfan syndrome has an estimated incidence in the general population of 2–3/10,000 (Pyeritz and McKusick 1979). Representing a single major diagnostic criteria, skeletal features can collectively include increased height and arm span, anterior chest deformity, joint laxity, pes planus, arachnodactyly, scoliosis, lordosis, dental crowding, and hammer toes. Craniofacial features associated with Marfan syndrome include malar hypoplasia, downslanting palpebral fissures, enophthalmos, and retrognathia

(De Paepe et al. 1996). Spontaneous pneumothorax can be an initial presenting feature of Marfan syndrome. Striae atrophicae and inguinal hernia represent minor features observed in individuals with Marfan syndrome. The prolonged survival of patients with Marfan syndrome, attributable to improved cardiac treatment, has revealed a number of non-cardiac health issues, including musculoskeletal concerns that may affect quality of life (Shores et al. 1994).

A deficiency of fibrillin-1 is postulated to alter the sequestration of the large latent TGF-β complex increasing its susceptibility for activation, thus contributing to the disease pathogenesis (Neptune et al. 2003). Fibrillin-1, a large molecular weight (350 kDa) glycoprotein, is an important component of the microfibrils which form the elastic fibers (Fig. 7.1).

Linkage analysis of *FBN1*, elastin, and *COL1A2* was performed in 11 families including 96 individuals in which adolescent idiopathic scoliosis segregated as an autosomal dominant condition (Miller et al. 1996). These loci were excluded as candidates for adolescent idiopathic scoliosis and were in agreement with prior data indicating that there was no deficiency in fibrillin incorporation in fibroblasts obtained from scoliotic patients.

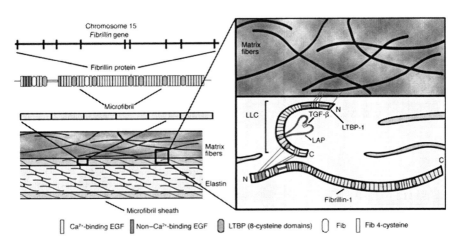

Fig. 7.1 Fibrillin-1 is translated from an mRNA encoded by the *FBN1* gene on chromosome 15. The protein is processed and secreted and then forms microfibrils in the matrix in association with other extracellular molecules. Microfibrils may function alone or form the basis of elastic fibers following deposition of elastin. Among molecules that are associated with fibrillin-1 is LTBP-1, a binding protein for the latent form of TGF-β. TGF-β is expressed as a precursor, which is cleaved to make the active form, but the two peptides remain associated and are then bound by a separate protein, LTBP-1. This complex, known as a large latent complex (LLC) is bound to elements of the extracellular matrix, including fibrillin-1 and other fibrils, and TGF-β is released in response to environmental and molecular signals. Expression of only half the normal amount of wild-type fibrillin-1 causes global impairment in the deposition of microfibrils. This could have dramatic effects on tissue structure and function (reprinted with permission: Byers (2004). Determination of the molecular basis of Marfan syndrome: a growth industry. J. Clin. Invest. 114:161–163)

Ehlers–Danlos Syndrome

The Ehlers–Danlos syndromes are a group of hereditary connective tissue disorders which are characterized by tissue fragility of the skin, ligaments, internal organs, and blood vessels. The estimated prevalence is approximately 1 in 5000 (Steinmann et al. 2002). Recent classification for Ehlers–Danlos syndrome has resulted in 10 clinical subtypes for it (Callewaert et al. 2008). Four subtypes in which scoliosis is an important clinical feature are described in Table 7.2, which lists some commonly encountered forms of Ehlers–Danlos syndrome for which scoliosis is a significant component.

Table 7.2 Some commonly encountered forms of Ehlers–Danlos syndrome associated with scoliosis

Clinical subtype	Clinical manifestations	Associated mutation	Biochemical consequence of gene defect
Classic (AD)	Skin hyperextensibility; wide, atrophic scar; easy bruising; molluscoid pseudotumors; complications of joint hypermobility	COL5A1, COL5A2 (Malfait et al. 2006)	Type V collagen functionally deficient. Type V collagen co-assembles with type I collagen and regulates collagen fibril diameter
"Classic-like" (AR)	Skin hyperextensibility; easy bruising; joint hypermobility; complications of joint hypermobility	Homozygous or compound heterozygous mutations in TNX-B (Schalkwijk et al. 2001)	Complete tenascin-X deficiency. Tenascin-X is a large extracellular matrix protein which is thought to play a role in collagen deposition by dermal fibroblasts
Hypermobility (AD)	Generalized joint hypermobility; mild skin involvement; multiple joint dislocations; chronic joint pain	TNX-B non functional allele; missense mutation in TNX-B; COL5A1 non-functional allele (Zweers et al. 2003)	~50% normal tenascin synthesized; tenascin-X structurally altered. Type V procollagen functionally deficient
Kyphoscoliotic (AR)	Kyphoscoliosis at birth; scleral fragility and rupture of globe; atrophic scar; easy bruising; Marfanoid habitus; arterial rupture	Homozygous or compound heterozygous mutations in LH-1 (PLOD1) (Yeowell and Walker 2000)	Deficiency of lysyl hydroxylase I which is required for hydroxylation of specific lysine residues to hydroxylysines

Given the important role the genes associated with these forms of Ehlers–Danlos syndrome play in the synthesis and maintenance of connective tissue, it is reasonable to postulate that they could contribute to the pathogenesis of idiopathic scoliosis.

Spondylocarpotarsal Syndrome

Spondylocarpotarsal syndrome is an autosomal recessive condition in which the salient features include short stature, vertebral, carpal, and tarsal fusions (Steiner et al. 2000, Honeywell et al. 2002). Nonsense mutations in filamin B (*FLNB*) have been associated with spondylocarpotarsal syndrome (Krakow et al. 2004). In addition, *FLNB* mutations have been associated with Larsen syndrome, atelosteogenesis I and III, and boomerang dysplasia (Krakow et al. 2004, Bicknell et al., 2005).

In mice, filamin B was present at the fetal femur growth plate. In proliferating chondrocytes, filamin B was concentrated at the cleavage region between dividing chondrocytes. Filamin B is also localized to the cartilaginous condensations of developing vertebrae. Filamins bind actin and stabilize the actin cytoskeleton, link the actin network with cellular membranes and mediate interactions between actin and transmembrane receptors (Stossel et al. 2001). They are dimerized structures which are associated at their carboxyl terminal region (Weihing 1985). The description of a case of an affected mother and son in the absence of a *FLNB* mutation provides evidence for an autosomal dominant form of this condition and genetic heterogeneity of this condition (Isidor et al. 2008). Presently no association between *FLNB* and idiopathic scoliosis has been shown, although given its presence in vertebral bodies the potential for this cannot be excluded.

Stickler Syndrome

Stickler syndrome, or hereditary arthro-ophthalmopathy, is an autosomal dominant connective tissue disorder which is associated with orofacial clefting, ocular, and musculoskeletal pathology. The condition occurs with a frequency of approximately 1 in 10,000 (Herrmann et al. 1975). Pierre Robin sequence or cleft palate may be present. Ocular findings include myopia which can progress to retinal detachment (Liberfarb et al. 2003). Type 1 Stickler syndrome is characterized by a vestigial gel remnant occupying the anterior vitreous cavity, which extends in a thin sheet over the pars plana and anterior retina (Snead et al. 1996a). Type 2 Stickler syndrome has abnormal vitreous architecture which is distributed throughout the posterior segment (Snead et al. 1996b). High frequency mixed and sensorineural hearing loss is a feature of Stickler syndrome (Liberfarb et al. 2003). Scoliosis occurs with a frequency of approximately 34% (Rose et al. 2001). Vertebral endplate abnormalities include sclerosis, Schmorl's nodes, platyspondyly, anterior spurring, Scheuermann-like kyphosis, disc space narrowing, and anterior cystic changes. Stickler syndrome type 1 is caused by mutations in *COL2A1* and type 2 by mutations in *COL11A1* and

COL11A2. The fibrillar collagen mutations in *COL2A1, COL11A1,* and *COL11A2* are thought to result in malformation and progressive weakening of intervertebral disks and vertebral endplates. Abnormal chondrification or endochondral ossification during development is hypothesized to result in vertebral anomalies, scoliosis, and kyphosis (Francomano et al. 1987).

Fibronectin

Fibronectin is an extracellular matrix glycoprotein, which is expressed at the sites of bone formation and is essential for osteoblast formation. Fibronectin polymerization mediated by direct interactions with integrins is necessary for collagen fibril assembly in the extracellular matrix (Kadler et al. 2008). Presently there is no role for fibronectin associated with the development of scoliosis in humans. Transgenic mice, in which fibronectin is deleted in the osteoblast lineage, exhibit vertebral deformities resembling scoliosis including fusions, hemivertebrae, block vertebrae and disorders of segmentation. Tail deformities ranged from involvement of a single vertebra to all tail vertebrae. Fibronectin deficiency was associated with decreased concentration of two extracellular matrix type proteins, including type I collagen and fibrillin-1 (Chen et al. 2008). In addition, decreased BMP signaling was observed in osteoblasts. Fibronectin is localized at somite boundaries. Potential roles for fibronectin in the development of scoliosis could include disruption of extracellular matrix formation and alterations in axial skeletal patterning, resulting in idiopathic and congenital scoliosis. Additional evidence in humans and animal models is needed to further evaluate this hypothesis.

Prader–Willi Syndrome

For the purpose of this review, Prader–Willi syndrome may be viewed as a prototypic chromosomal syndrome where scoliosis is a significant component. Prader–Willi syndrome, occurring with a frequency of 1 in 25,000 live births, is characterized by neonatal hypotonia, failure to thrive, hyperphagia, morbid obesity, short stature, learning, and behavior problems (Prader et al. 1956). Diagnostic criteria for Prader–Willi syndrome have been established by Holm et al. (1993). In approximately 75% of affected individuals, Prader–Willi syndrome is caused by a small deletion in paternally derived chromosome 15q11–15q13. Approximately 20% of affected individuals with Prader–Willi syndrome have uniparental disomy for chromosome 15 and less than 5% of affected individuals have an imprinting defect in the imprinting center localized within 15q11–q13, also referred to as the Prader–Willi/Angelman syndrome region since Prader–Willi syndrome and Angelman syndrome share the same chromosome 15q11 region. The genetics of Prader–Willi syndrome is complex and the reader may be referred to more detailed reviews of the subject (Cassidy and Driscoll 2009).

Scoliosis is observed with an estimated overall frequency of 43.4% in Prader–Willi syndrome with a prevalence of 70% at skeletal maturity (Odent et al. 2008). Twenty-four of the 63 scoliotic patients required bracing and 16 patients required surgical treatment. The development of scoliosis in Prader–Willi syndrome appears to have a multifactorial causation with risk factors including hypotonia of paravertebral musculature, obesity, and growth hormone treatment. There are several genes which are usually expressed on the paternally derived chromosome 15, which are inactivated in patients with Prader–Willi syndrome. *SNRPN* (small nuclear ribonucleoprotein N) is expressed predominately in the brain (Özçelik et al. 1992). The remaining genes in the Prader–Willi syndrome/Angelman syndrome region include *MKRN3, MAGEL2,* and *NECDIN.* In addition, there are five repetitious snoRNA genes (*HBII-436, HBII-13, HBII-438, HBII-85,* and *HBII-52*) and the antisense transcript to *UBE3A*, a gene associated with major phenotypic features in Angelman syndrome. It is uncertain how these gene(s) may contribute to the pathogenesis of scoliosis.

Neuromuscular and Neurogenic Etiologies

Scoliosis is a feature associated with muscular dystrophies (Do 2002). The genetic etiologies are heterogeneous and reflect the metabolism and architecture of the muscle unit. Candidate gene approaches have been utilized to identify genes associated with muscle development which could potentially contribute to idiopathic scoliosis. One such example is *SNTG1*. This gene encodes γ1-syntropin, a dystrophin-associated protein occurring in muscle cells and in Purkinje neurons in the cerebellum, and has been studied as a candidate gene in a family in which a pericentric inversion involving chromosome 8 segregates with idiopathic scoliosis. A 6-bp deletion immediately adjacent to the 3' exon 10 was observed in 5 of 7 affected individuals in this family and 0 of 480 control chromosomes. Despite the discordant segregation of this mutation with affected family members, *SNTG1* remains a candidate gene for idiopathic scoliosis (Bashiardes et al. 2004).

Using limb girdle muscular dystrophy 2A, an autosomal recessive disorder caused by mutations in calpain 3 (*CAPN3*), a calcium-dependent cystein muscle-specific protease, as a model, RNA expression profiling in muscle biopsies from affected patients was compared to controls in order to identify genes with altered expression (Sáenz et al. 2008). Genes associated with extracellular matrix development including collagen type I and III and cell adhesion proteins such as fibronectin, CD9, and CD44 were upregulated. This pattern of increased expression is due to fibrotic infiltration of muscle cells. Several genes associated with muscle development were upregulated. These include (1) myosin, heavy chain 3, skeletal muscle, embryonic (*MYH3*), (2) myosin light chain 1 slow A (*MLC1SA*), a transcriptional regulator, which is expressed in slow-twitch skeletal muscle and non-muscle tissue and promotes muscle cell proliferation, (3) insulin-like growth factor (IGF)-1, which has been implicated in gene expression control in cell differentiation, proliferation, and degradation. IGF-1 also has anabolic effects on muscle cells mediated by its receptor, a tyrosine kinase activated upon IGF-1 binding ultimately resulting in increased protein synthesis (Baldwin and Haddad 2002, Marotta et al. 2007). The upregulation of frizzled-related protein (FRZB) suggests altered β-catenin regulation is altered at the level of the Wnt signaling pathway, resulting in alteration in myogenesis. Although these expression alterations were observed in muscle cells, it remains open to further investigation as to what, if any, effects these changes have on spinal curvature in the patient with idiopathic scoliosis.

Smith–Magenis Syndrome

Scoliosis may occur in Smith–Magenis syndrome due to alterations in vertebral morphogenesis or other mechanisms that may alter spinal curvature.

Smith–Magenis syndrome is a contiguous gene disorder which results from an interstitial deletion of chromosome 17p11.2 and is associated with a characteristic facial appearance including brachycephaly, broadening of the jaw, tenting of the lips, downturned mouth, and mid-facial hypoplasia (Fig. 7.2a). Additional features include a high-pitched voice and pes planus or pes cavus (Greenberg et al. 1991, Greenberg et al. 1996). Children with Smith–Magenis syndrome exhibit

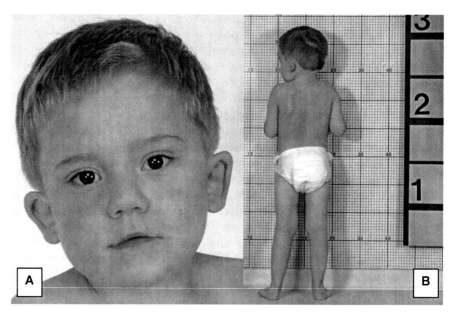

Fig. 7.2 (**a**) Male child with Smith–Magenis syndrome. Facial features include mild mid-facial hypoplasia, deeply set, upslanting palpebral fissures, tenting of the upper lip. (**b**) Scoliosis is evident on posterior view. Photographs courtesy of Ann C.M. Smith M.A. D.Sc. (hon), NHGRI/NIH

developmental delay/mental retardation and unique complex phenotypic behaviors characterized by nail yanking and polyembolokoilamania or insertions of objects into body orifices (Smith et al. 1998). Other behavioral problems include impulsivity, aggression, hyperactivity, distractibility, and sleep disturbances (Smith et al. 1998). It is noteworthy that scoliosis occurs with a frequency of approximately 60% in patients with Smith–Magenis syndrome (Greenberg et al. 1996). Some patients with Smith–Magenis syndrome display congenital scoliosis (Smith et al. 1986) (Fig. 7.2b).

While 17p11.2 microdeletions account for the majority of cases with Smith–Magenis syndrome, 5–10% of patients with Smith–Magenis syndrome have mutations in the *RAI1* gene (Slager et al. 2003). Presently the role of *RAI1* is unclear, although it is thought to contribute to neuronal differentiation. The RAI1 protein is thought to localize to the nucleus and stimulate transcription through the interaction with other DNA-binding proteins (Slager et al. 2003).

In *Rai1$^{+/-}$* targeted null allele mice, *Rai1* expression was observed in branchial arches, forelimb bud, and hindlimb bud (Bi et al.2005). SDVs, including failure of closure of neural arches of C5 and C7, and absence of the spinal process in T2 were observed. The transverse foramina in C4, C5, and C6 vertebrae were not closed. In the brain of *Rai1$^{+/-}$* targeted null allele mice, *Rai1* was largely expressed in the hippocampus and cerebellum (Bi et al. 2005). Retinoic acid plays an important role in development of the spinal column. Opposing gradients of retinoic acid and fibroblast growth factor (FGF) are necessary for somitogenesis (Moreno and Kitner 2004).

Individuals with Smith–Magenis syndrome have sleep disturbances starting as early as infancy (Greenberg et al. 1996). This is thought to be due to an inverted pattern of melatonin secretion resulting in abnormalities in the production, secretion, distribution, and metabolism of melatonin (Potocki et al. 2000). Alterations in melatonin signaling dysfunction have been hypothesized to contribute to the development of idiopathic scoliosis (Moreau et al. 2004). This has been supported by observations that pinealectomy in chicken and rats maintained in a bipedal mode results in scoliosis (McCarrey et al. 1981, Muccielli et al. 1996, Wang et al. 1998, Machida et al. 1999). However, circulating levels of melatonin in humans with adolescent idiopathic scoliosis are within the normal range (Wang et al. 1998, Bagnall et al 1999). *In vivo* assays of melatonin signaling performed on osteoblasts obtained from adolescents with idiopathic scoliosis demonstrated the presence of three groups characterized by varying degrees of impaired responsiveness to melatonin (Moreau et al. 2004). This suggests that three distinct mutations could cause adolescent idiopathic scoliosis. In subsequent studies, G inhibitory proteins associated with melatonin cell surface receptors were hypofunctional due to increased phosphorylation of serine residues Azeddine et al. 2007. Figure 7.3a–d illustrates a hypothesis for altered melatonin signaling in the pathogenesis of adolescent idiopathic scoliosis. The association of sleep disturbances involving a neurohormonal deficit or dysfunction reported in adolescent idiopathic scoliosis provides additional evidence supporting a role for altered melatonin signal dysfunction in adolescent idiopathic scoliosis.

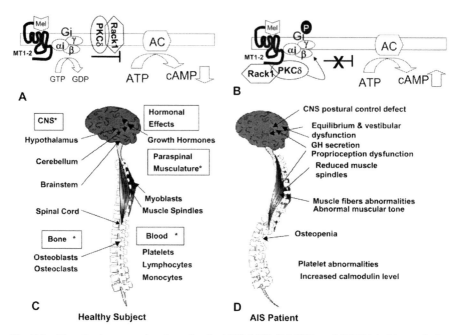

Fig. 7.3 Hypothesis concerning the role of cAMP, MT2, RACK1, and PKC[delta] in scoliotic and non-scoliotic tissues is proposed here. Normal melatonin signaling pathway is summarized in (**a**) whereas (**b**) illustrates a model showing complex interactions interfering with the transmission of melatonin signal. In control cells, melatonin binds to its receptor and activates Gi alpha subunit to inhibit adenylyl cyclase activity. In AIS patients, Gi proteins are inactivated through phosphorylation by PKC[delta] and/or other kinases. Second, PKC[delta] interaction with MT2 melatonin receptor could lead to its phosphorylation and to desensitization of MT2 receptor. Melatonin signaling occurs in different systems and tissues (**c**). The potential systemic defects resulting from a melatonin signaling dysfunction in adolescent idiopathic scoliosis could be responsible for many physiological defects. This proposed model of adolescent idiopathic scoliosis could explain the multiplicity of disorders reported in AIS pathogenesis (**d**). In addition, sleep disturbances involving a neurohormonal deficit or dysfunction have been often reported in adolescent idiopathic scoliosis and other neuromuscular disorders suggesting the involvement of melatonin (Reprinted with permission: Azeddine et al. (2007). Molecular determinants of melatonin signaling dysfunction in adolescent idiopathic scoliosis. Clin. Orthop. Relat. Res. 462:45–52)

These observations suggest that haploinsufficiency of *RAI1* in humans can contribute to the occurrence of congenital and idiopathic scoliosis through alteration in vertebral morphogenesis in addition to alteration in spinal curvature later in childhood. Thus, scoliosis may be central nervous system mediated.

Concluding Remarks

This chapter has focused on genetic etiologies for congenital, neuromuscular, and idiopathic scoliosis. Emphasis was placed on a multifactorial causation for idiopathic scoliosis which involves interplay of developmental, neuromuscular, and connective tissue factors which could synergistically predispose to spinal curvature. Proof of these hypotheses awaits results derived from further human and animal studies.

Acknowledgments The author thanks Marshfield Clinic Research Foundation for its support through the assistance of Carol Beyer, Marie Fleisner, Dr. Ingrid Glurich, and Alice Stargardt in the preparation of this chapter.

References

Abe, K., Yamamura, K., and Suzuki, M., 2000. Molecular and embryological characterization of a new transgene-induced null allele of mouse Brachyury locus. Mamm. Genome 11:238–240.

Alagille, D., Odièvre, M., Gautier, M., and Dommergues, J.P., 1975. Hepatic ductular hypoplasia associated with characteristic facies, vertebral malformations, retarded physical, mental, and sexual development, and cardiac murmur. J. Pediatr. 86:63–71.

Azeddine, B., Letellier, K., Wang da, S., Moldovan, F., and Moreau, A., 2007. Molecular determinants of melatonin signaling dysfunction in adolescent idiopathic scoliosis. Clin. Orthop. Relat. Res. 462:45–52.

Bagnall, K., Raso, V.J., Moreau, M., Mahood, J., Wang, X., and Zhao, J. 1999. The effects of melatonin therapy on the development of scoliosis after pinealectomy in the chicken. J. Bone. Joint. Surg. Am. 81:91–99.

Baldwin, K.M., and Haddad, F. 2002. Skeletal muscle plasticity: cellular and molecular responses to altered physical activity paradigms. Am. J. Phys. Med. Rehabil. 81(11 Suppl):S40–S51.

Band, M.R., Olmstead, C., Everts, R.E., Liu, Z.L., and Lewin, H.A., 2002. A 3800 gene microarray for cattle functional genomics: comparison of gene expression in spleen, placenta, and brain. Anim. Biotechnol. 13:163–172.

Bashiardes, S., Veile, R., Allen, M., Wise, C.A., Dobbs, M., Morcuende, J.A., Szappanos, L., Herring, J.A., Bowcock, A.M., and Lovett, M. 2004. SNTG1, the gene encoding gamma1-syntrophin: a candidate gene for idiopathic scoliosis. Hum. Genet. 115:81–89.

Belloni, E., Martucciello, G., Verderio, D., Ponti, E., Seri, M., Jasonni, V., Torre, M., Ferrari, M., Tsui, L.C., and Scherer, S.W. 2000. Involvement of the HLXB9 homeobox gene in Currarino syndrome. Am. J. Hum. Genet. 66:312–319.

Bengtsson, B.E., Larsson, A., Bengtsson, A., and Renberg, L. 1988. Sublethal effects of tetrachloro-1,2-benzoquinone–a component in bleachery effluents from pulp mills–on vertebral quality and physiological parameters in fourhorn sculpin. Ecotoxicol. Environ. Saf. 15:62–71.

Bi, W., Ohyama, T., Nakamura, H., Yan, J., Visvanathan, J., Justice, M.J., and Lupski, J.R. 2005. Inactivation of Rai1 in mice recapitulates phenotypes observed in chromosome engineered mouse models for Smith-Magenis syndrome. Hum. Mol. Genet. 14:983–995.

Bicknell, L.S., Morgan, T., Bonafé, L., Wessels, M.W., Bialer, M.G., Willems, P.J., Cohn, D.H., Krakow, D., and Robertson, S.P. 2005. Mutations in FLNB cause boomerang dysplasia. J. Med. Genet. 42:e43.

Blumel, J., Evans, E.B., and Eggers, G.W. 1959. Partial and complete agenesis or malformation of the sacrum with associated anomalies; etiologic and clinical study with special reference to heredity; a preliminary report. J. Bone Joint Surg. Am. 41A:497–518.

Bogani, D., Warr, N., Elms, P., Davies, J., Tymowska-Lalanne, Z., Goldsworthy, M., Cox, R.D., Keays, D.A., Flint, J., Wilson, V., Nolan, P., and Arkell, R. 2004. New semidominant mutations that affect mouse development. Genesis 40:109–117.

Botto, L.D., Khoury, M.J., Mastroiacovo, P., Castilla, E.E., Moore, C.A., Skjaerven, R., Mutchinick, O.M., Borman, B., Cocchi, G., Czeizel, A.E., Goujard, J., Irgens, L.M., Lancaster, P.A., Martínez-Frías, M.L., Merlob, P., Ruusinen, A., Stoll, C., and Sumiyoshi, Y. 1997. The spectrum of congenital anomalies of the VATER association: an international study. Am. J. Med. Genet. 71:8–15.

Burck, U. 1983. Genetic aspects of hemifacial microsomia. Hum. Genet. 64, 291–296.

Callewaert, B., Malfait, F., Loeys, B., De Paepe, A., 2008. Ehlers-Danlos syndromes and Marfan syndrome. Best Pract. Res. Clin. Rheumatol. 22:165–189.

Byers, P.H. 2004. Determination of the molecular basis of Marfan syndrome: a growth industry. J. Clin. Invest. 114:161–163.

Cassidy, S.B., and Driscoll, D.J., 2009. Prader-Willi syndrome. Eur. J. Hum. Genet. 17:3–13.

Chen, Q., Zhao, H., Zhao, J., Pacicca, D.M., Fassler, R., and Dallas, S.L. 2008. Conditional deletion of fibronectin results in a scoliosis-like phenotype. ASBMR 23:S15.

Chesley, P. 1935. Development of the short-tailed mutant in the house mouse. J. Exp. Zool. 70: 429–459.

Clarke, R.A., Singh, S., McKenzie, H., Kearsley, J.H., and Yip, M.Y. 1995. Familial Klippel-Feil syndrome and paracentric inversion inv(8)(q22.2q23.3). Am. J. Hum. Genet. 57, 1364–1370.

Clarke, R.A., Catalan, G., Diwan, A.D., and Kearsley, J.H. 1998. Heterogeneity in Klippel-Feil syndrome: a new classification. Pediatr. Radiol. 28:967–974.

Cohen, J. 2007. Genomics. DNA duplications and deletions help determine health. Science 317:1315–1317.

Colliton, R.P., Bason, L., Lu, F.M., Piccoli, D.A., Krantz, I.D., and Spinner, N.B. 2001. Mutation analysis of Jagged1 (JAG1) in Alagille syndrome patients. Hum. Mutat. 17: 151–152.

Crétolle, C., Pelet, A., Sanlaville, D., Zérah, M., Amiel, J., Jaubert, F., Révillon, Y., Baala, L., Munnich, A., Nihoul-Fékété, C., and Lyonnet, S. 2008. Spectrum of HLXB9 gene mutations in Currarino syndrome and genotype-phenotype correlation. Hum. Mutat. 29:903–910.

Currarino, G., Coln, D., and Votteler, T. 1981. Triad of anorectal, sacral, and presacral anomalies. Am. J. Roentgenol. 137:395–398.

De Paepe, A., Devereux, R.B., Dietz, H.C., Hennekam, R.C., and Pyeritz, R.E. 1996. Revised diagnostic criteria for the Marfan syndrome. Am. J. Med. Genet. 62:417–426.

Dietrich, S., and Gruss, P. 1995. Undulated phenotypes suggest a role of Pax-1 for the development of vertebral and extravertebral structures. Dev. Biol. 167:529–548.

Dietz, H.C., Cutting, G.R., Pyeritz, R.E., Maslen, C.L., Sakai, L.Y., Corson, G.M., Puffenberger, E.G., Hamosh, A., Nanthakumar, E.J., Curristin, S.M., Stetten, G., Meyers, D.A., and Francomano, C.A. 1991. Marfan syndrome caused by a recurrent de novo missense mutation in the fibrillin gene. Nature 352:337–339.

Do, T. 2002. Orthopedic management of the muscular dystrophies. Curr. Opin. Pediatr. 14:50–53.

Edwards, M.J. 1986. Hyperthermia as a teratogen: a review of experimental studies and their clinical significance. Teratog. Carcinog. Mutagen. 6:563–582.

Emerick, K.M., Rand, E.B., Goldmuntz, E., Krantz, I.D., Spinner, N.B., and Piccoli, D.A. 1999. Features of Alagille syndrome in 92 patients: frequency and relation to prognosis. Hepatology 29:822–829.

Ewart-Toland, A., Yankowitz, J., Winder, A., Imagire, R., Cox, V.A., Aylsworth, A.S., and Golabi, M. 2000. Oculoauriculovertebral abnormalities in children of diabetic mothers. Am. J. Med. Genet. 90:303–309.

Farley, F.A., Loder, R.T., Nolan, B.T., Dillon, M.T., Frankenburg, E.P., Kaciroti, N.A., Miller, J.D., Goldstein, S.A., and Hensinger, R.N. 2001. Mouse model for thoracic congenital scoliosis. J. Pediatr. Orthop. 21:537–540.

Feil, A. 1919. L'absence et la diminution Des vertebres cervicales. Thesis, Libraire Litteraire et Medicale, Paris.

Fischer, S., Lüdecke, H.J., Wieczorek, D., Böhringer, S., Gillessen-Kaesbach, G., and Horsthemke, B. 2006. Histone acetylation dependent allelic expression imbalance of BAPX1 in patients with the oculo-auriculo-vertebral spectrum. Hum. Mol. Genet. 15:581–587.

Francomano, C.A., Liberfarb, R.M., Hirose, T., Maumenee, I.H., Streeten, E.A., Meyers, D.A., and Pyeritz, R.E. 1987. The Stickler syndrome: evidence for close linkage to the structural gene for type II collagen. Genomics 1:293–296.

Fukushima, Y., Ohashi, H., Wakui, K., Nishimoto, H., Sato, M., and Aihara, T. 1995. De novo apparently balanced reciprocal translocation between 5q11.2 and 17q23 associated with Klippel-Feil anomaly and type A1 brachydactyly. Am. J. Med. Genet. 57, 447–449.

Garcia-Barcelo, M.M., Wong, K.K., Lui, V.C., Yuan, Z.W., So, M.T., Ngan, E.S., Miao, X.P., Chung, P.H., Khong, P.L., and Tam, P.K. 2008. Identification of a *HOXD13* mutation in a VACTERL patient. Am. J. Med. Genet. A 146A:3181–3185.

Ghebranious, N., Blank, R.D., Raggio, C.L., Staubli, J., McPherson, E., Ivacic, L., Rasmussen, K., Jacobsen, F.S., Faciszewski, T., Burmester, J.K., Pauli, R.M., Boachie-Adjei, O., Glurich, I., and Giampietro, P.F. 2008. A missense T(Brachyury) mutation contributes to vertebral malformations. J. Bone Miner. Res. 23:1576–1583.

Ghebranious, N., Burmester, J.K., Glurich, I., McPherson, E., Ivacic, L., Kislow, J., Rasmussen, K., Kumar, V., Raggio, C.L., Blank, R.D., Jacobsen, F.S., Faciszewski, T., Womack, J., and Giampietro, P.F. 2006. Evaluation of SLC35A3 as a candidate gene for human vertebral malformations. Am. J. Med. Genet. 140:1346–1348.

Ghebranious, N., Raggio, C.L., Blank, R.D., McPherson, E., Burmester, J.K., Ivacic, L., Rasmussen, K., Kislow, J., Glurich, I., Jacobsen, F.S., Faciszewski, T., Pauli, R.M., Boachie-Adjei, O., and Giampietro, P.F. 2007. Lack of evidence of WNT3A as a candidate gene for congenital vertebral malformations. Scoliosis 2:13.

Giampietro, P.F., Blank, R.D., Raggio, C.L., Merchant, S., Jacobsen, F.S., Faciszewski, T., Shukla, S.K., Greenlee, A.R., Reynolds, C., and Schowalter, D.B. 2003. Congenital and idiopathic scoliosis: clinical and genetic aspects. Clin. Med. Res. 1:125–136.

Giampietro, P.F., Raggio, C.L., and Blank, R.D. 1998. Synteny-defined candidate genes for congenital and idiopathic scoliosis. Am. J. Med. Genet. 83:164–177.

Giampietro, P.F., Raggio, C.L., Reynolds, C., Ghebranious, N., Burmester, J.K., Glurich, I., Rasmussen, K., McPherson, E., Pauli, R.M., Shukla, S.K., Merchant, S., Jacobsen, F.S., Faciszewski, T., and Blank, R.D. 2006. DLL3 as a candidate gene for vertebral malformations. Am. J. Med. Genet. A 140:2447–2453.

Giampietro, P.F., Raggio, C.L., Reynolds, C.E., Shukla, S.K., McPherson, E., Ghebranious, N., Jacobsen, F.S., Kumar, V., Faciszewski, T., Pauli, R.M., Rasmussen, K., Burmester, J.K., Zaleski, C., Merchant, S., David, D., Weber, J.L., Glurich, I., and Blank, R.D. 2005. An analysis of PAX1 in the development of vertebral malformations. Clin. Genet. 68: 448–453.

Giampietro, P.F., Dunwoodie, S.L., Kusumi, K., Pourquié, O., Tassy, O., Offiah, A.C., Cornier, A.S., Alman, B.A., Blank, R.D., Raggio, C.L., Glurich, I., and Turnpenny, P.D. 2009. Molecular diagnosis of vertebral segmentation disorders in humans. Expert Opin. Med. Diagn. 2: 1107–1121

Goto, M., Nishimura, G., Nagai, T., Yamazawa, K., and Ogata, T. 2006. Familial Klippel-Feil anomaly and t(5;8)(q35.1;p21.1) translocation. Am. J. Med. Genet. A 140, 1013–1015.

Greenberg, F., Guzzetta, V., Montes de Oca-Luna, R., Magenis, R.E., Smith, A.C., Richter, S.F., Kondo, I., Dobyns, W.B., Patel, P.I., and Lupski, J.R. 1991. Molecular analysis of the Smith-Magenis syndrome: a possible contiguous-gene syndrome associated with del(17)(p11.2). Am. J. Hum. Genet. 49:1207–1218.

Greenberg, F., Lewis, R.A., Potocki, L., Glaze, D., Parke, J., Killian, J., Murphy, M.A., Williamson, D., Brown, F., Dutton, R., McCluggage, C., Friedman, E., Sulek, M., and Lupski, J.R. 1996. Multi-disciplinary clinical study of Smith-Magenis syndrome (deletion 17p11.2). Am. J. Med. Genet. 62:247–254.

Hanold, K.C. 1986. Teratogenic potential of valproic acid. J. Obstet. Gynecol. Neonatal Nurs. 15:111–116.

Harlow, C.L., Partington, M.D., and Thieme, G.A. 1995. Lumbosacral agenesis: clinical characteristics, imaging, and embryogenesis. Pediatr. Neurosurg. 23:140–147.

Harrison, S.M., Houzelstein, D., Dunwoodie, S.L., and Beddington, R.S. 2000. Sp5, a new member of the Sp1 family, is dynamically expressed during development and genetically interacts with Brachyury. Dev. Biol. 227:358–372.

Hensinger, R.N., and MacEwen, C.D. 1982. In Congenital Anomalies of the Spine, eds. R. Rothman and F. Simeone, pp. 216–233. Philadelphia: W.B. Saunders

Herrmann, J., France, T.D., Spranger, J.W., Opitz, J.M., and Wiffler, C. 1975. The Stickler syndrome (hereditary arthroophthalmopathy). Birth Defects Orig. Artic. Ser. 11, 76–103.

Hickory, W., Nanda, R., and Catalanotto, F.A. 1979. Fetal skeletal malformations associated with moderate zinc deficiency during pregnancy. J. Nutr. 109:883–891.

Hingorani, M., Nischal, K.K., Davies, A., Bentley, C., Vivian, A., Baker, A.J., Mieli-Vergani, G., Bird, A.C., and Aclimandos, W.A. 1999. Ocular abnormalities in Alagille syndrome. Ophthalmology 106:330–337.

Holm, V.A., Cassidy, S.B., Butler, M.G., Hanchett, J.M., Greenswag, L.R., Whitman, B.Y., and Greenberg, F. 1993. Prader-Willi syndrome: consensus diagnostic criteria. Pediatrics 91: 398–402.

Honeywell, C., Langer, L., and Allanson, J. 2002. Spondylocarpotarsal synostosis with epiphyseal dysplasia. Am. J. Med. Genet. 109:318–322.

Hutchinson, J. 1894. Deformity of shoulder girdle. Br. Med. J. 1:634–635.

Isidor, B., Cormier-Daire, V., Le Merrer, M., Lefrancois, T., Hamel, A., Le Caignec, C., David, A., and Jacquemont, S. 2008. Autosomal dominant spondylocarpotarsal synostosis syndrome: phenotypic homogeneity and genetic heterogeneity. Am. J. Med. Genet. A 146A: 1593–1597.

Jaskwhich, D., Ali, R.M., Patel, T.C., and Green, D.W. 2000. Congenital scoliosis. Curr. Opin. Pediatr. 12:61–66.

Jensen, L.E., Barbaux, S., Hoess, K., Fraterman, S., Whitehead, A.S., and Mitchell, L.E. 2004. The human T locus and spina bifida risk. Hum. Genet. 115:475–482.

Kadler, K.E., Hill, A., Canty-Laird, E.G., 2008. Collagen fibrillogenesis: fibronectin, integrins, and minor collagens as organizers and nucleators. Curr. Opin. Cell Biol. 20:495–501.

Kamath, B.M., Loomes, K.M., Oakey, R.J., Emerick, K.E., Conversano, T., Spinner, N.B., Piccoli, D.A., and Krantz, I.D. 2002. Facial features in Alagille syndrome: specific or cholestasis facies? Am. J. Med. Genet. 112:163–170.

Katz, L.A., Schultz, R.E., Semina, E.V., Torfs, C.P., Krahn, K.N., and Murray, J.C. 2004. Mutations in PITX2 may contribute to cases of omphalocele and VATER-like syndromes. Am. J. Med. Genet. A 130A:277–283.

Kispert, A., and Herrmann, B.G. 1994. Immunohistochemical analysis of the Brachyury protein in wild-type and mutant mouse embryos. Dev. Biol. 161:179–193.

Klippel, M., and Feil, A. 1912. Un cas d'absence des vertebres cervicales. Nouv. Iconog. Salpetriere. 25:223–250.

Knudson, A.G. Jr. 1971. Mutation and cancer: statistical study of retinoblastoma. Proc. Natl. Acad. Sci. 68:820–823.

Krakow, D., Robertson, S.P., King, L.M., Morgan, T., Sebald, E.T., Bertolotto, C., Wachsmann-Hogiu, S., Acuna, D., Shapiro, S.S., Takafuta, T., Aftimos, S., Kim, C.A., Firth, H., Steiner, C.E., Cormier-Daire, V., Superti-Furga, A., Bonafe, L., Graham, J.M., Jr., Grix, A., Bacino, C.A., Allanson, J., Bialer, M.G., Lachman, R.S., Rimoin, D.L., and Cohn, D.H. 2004. Mutations in the gene encoding filamin B disrupt vertebral segmentation, joint formation and skeletogenesis. Nat. Genet. 36:5–10.

Ladi, E., Nichols, J.T., Ge, W., Miyamoto, A., Yao, C., Yang, L.T., Boulter, J., Sun, Y.C., Kitner, C., and Weinmaster G. 2005. The divergent DSL ligand Dll3 does not activate Notch signaling but cell autonomously attenuates signaling induced by other DSL ligands. J. Cell Bio. 170: 983–992.

Lebepe-Mazur, S., Bal, H., Hopmans, E., Murphy, P., and Hendrich, S. 1995. Fumonisin B1 is fetotoxic in rats. Vet. Hum. Toxicol. 37:126–130.

Lee, S.K., Jurata, L.W., Funahashi, J., Ruiz, E.C., and Pfaff, S.L. 2004. Analysis of embryonic motoneuron gene regulation: derepression of general activators function in concert with enhancer factors. Development 131:3295–3306.

Lee, S.K., and Pfaff, S.L. 2003. Synchronization of neurogenesis and motor neuron speci-fication by direct coupling of bHLH and homeodomain transcription factors. Neuron 38: 731–745.

Lewis, D.P., Van Dyke, D.C., Stumbo, P.J., and Berg, M.J. 1998. Drug and environmental fac-tors associated with adverse pregnancy outcomes. Part II: improvement with folic acid. Ann. Pharmacother. 32:947–961.

Liberfarb, R.M., Levy, H.P., Rose, P.S., Wilkin, D.J., Davis, J., Balog, J.Z., Griffith, A.J., Szymko-Bennett, Y.M., Johnston, J.J., Francomano, C.A., Tsilou, E., and Rubin, B.I. 2003. The Stickler syndrome: genotype/phenotype correlation in 10 families with Stickler syndrome resulting from seven mutations in the type II collagen gene locus COL2A1. Genet. Med. 5:21–27.

Lyon, M.F. 1996. An additional type of male sterility and inherited urinary obstruction in mice with the t-haplotype th7. Genet. Res. 67:249–256.

Machida, M., Murai, I., Miyashita, Y., Dubousset, J., Yamada, T., and Kimura, J. 1999. Pathogenesis of idiopathic scoliosis. Experimental study in rats. Spine 24:1985–1989.

Malfait, F., Hakim, A.J., De Paepe, A., and Grahame, R. 2006. The genetic basis of the joint hypermobility syndromes. Rheumatology (Oxford) 45:502–507.

Manaligod, J.M., Bauman, N.M., Menezes, A.H., and Smith, R.J. 1999. Cervical vertebral anomalies in patients with anomalies of the head and neck. Ann. Otol. Rhinol. Laryneol. 108:925–933.

Marotta, M., Sarria, Y., Ruiz-Roig, C., Munell, F., and Roig-Quilis, M. 2007. Laser microdissection-based expression analysis of key genes involved in muscle regeneration in mdx mice. Neuromuscul. Disord. 17:707–718.

McCarrey, J.R., Abbott, U.K., Benson, D.R., and Riggins, R.S. 1981. Genetics of scoliosis in chickens. J. Hered. 72:6–10.

McDaniell, R., Warthen, D.M., Sanchez-Lara, P.A., Pai, A., Krantz, I.D., Piccoli, D.A., and Spinner, N.B. 2006. NOTCH2 mutations cause Alagille syndrome, a heterogeneous disorder of the notch signaling pathway. Am. J. Hum. Genet. 79:169–173.

McGaughran, J.M., Oates, A., Donnai, D., Read, A.P., and Tassabehji, M. 2003. Mutations in PAX1 may be associated with Klippel-Feil syndrome. Eur. J. Hum. Genet. 11, 468–474.

McMaster, M., and Ohtsuka, K. 1982. The natural history of congenital scoliosis. A study of two hundred and fifty-one patients. J. Bone Joint Surg. Am. 64:1128–1147.

McMaster, M.J. and Singh, H. 1999. Natural history of congenital kyphosis and kyphoscoliosis. A study of one hundred and twelve patients. J. Bone Joint Surg. Am. 1999 81:1367–1383.

Merello, E., De Marco, P., Mascelli, S., Raso, A., Calevo, M.G., Torre, M., Cama, A., Lerone, M., Martucciello, G., and Capra, V. 2006. HLXB9 homeobox gene and caudal regression syndrome. Birth Defects Res. A Clin. Mol. Teratol. 76:205–209.

Miller, N.H., Mims, B., Child, A., Milewicz, D.M., Sponseller, P., and Blanton, S.H. 1996. Genetic analysis of structural elastic fiber and collagen genes in familial adolescent idiopathic scoliosis. J. Orthop. Res. 14:994–999.

Moreau, A., Wang, D.S., Forget, S., Azeddine, B., Angeloni, D., Fraschini, F., Labelle, H., Poitras, B., Rivard, C.H., and Grimard, G. 2004. Melatonin signaling dysfunction in adolescent idiopathic scoliosis. Spine 29:1772–1781.

Moreno, T.A., and Kitner, C. 2004. Regulation of segmental patterning by retinoic acid signaling during Xenopus somitogenesis. Dev. Cell 6:205–218.

Morrison, K., Papapetrou, C., Attwood, J., Hol, F., Lynch, S.A., Sampath, A., Hamel, B., Burn, J., Sowden, J., Stott, D., Mariman, E., and Edwards, Y.H. 1996. Genetic mapping of the human homologue (T) of mouse T(Brachyury) and a search for allele association between human T and spina bifida. Hum. Mol. Genet. 5:669–674.

Muccielli, M.L., Martinez, S., Pattyn, A., Goridis, C., and Brunet, J.F. 1996. Otlx2, an Otx-related homeobox gene expressed in the pituitary gland and in a restricted pattern in the forebrain. Mol. Cell. Neurosci. 8:258–271.

Nakano, T., Windrem, M., Zappavigna, V., and Goldman, S.A. 2005. Identification of a conserved 125 base-pair Hb9 enhancer that specifies gene expression to spinal motor neurons. Dev. Biol. 283:474–485.

Neptune, E.R., Frischmeyer, P.A., Arking, D.E., Myers, L., Bunton, T.E., Gayraud, B., Ramirez, F., Sakai, L.Y., and Dietz, H.C. 2003. Dysregulation of TGF-beta activation contributes to pathogenesis in Marfan syndrome. Nat. Genet. 33:407–411.

Odent, T., Accadbled, F., Koureas, G., Cournot, M., Moine, A., Diene, G., Molinas, C., Pinto, G., Tauber, M., Gomes, B., de Gauzy, J.S., and Glorion, C. 2008. Scoliosis in patients with Prader-Willi syndrome. Pediatrics 122:e499–503.

Ozçelik, T., Leff, S., Robinson, W., Donlon, T., Lalande, M., Sanjines, E., Schinzel, A., and Francke, U. 1992. Small nuclear ribonucleoprotein polypeptide N (SNRPN), an expressed gene in the Prader-Willi syndrome critical region. Nat. Genet. 2:265–269.

Pang, D. 1993. Sacral agenesis and caudal spinal cord malformations. Neurosurgery 32: 755–778.

Papagrigorakis, M.J., Synodinos, P.N., Daliouris, C.P., and Metaxotou, C. 2003. De novo inv(2)(p12q34) associated with Klippel-Feil anomaly and hypodontia. Eur. J. Pediatr. 162: 594–597.

Papapetrou, C., Drummond, F., Reardon, W., Winter, R., Spitz, L., and Edwards, Y.H. 1999. A genetic study of the human T gene and its exclusion as a major candidate gene for sacral agenesis with anorectal atresia. J. Med. Genet. 36:208–213.

Park, C.H., Pruitt, J.H., and Bennett, D. 1989. A mouse model for neural tube defects: the curtailed (Tc) mutation produces spina bifida occulta in Tc/+ animals and spina bifida with meningomyelocele in Tc/t. Teratology 39:303–312.

Phelan, M.C., Rogers, R.C., Clarkson, K.B., Bowyer, F.P., Levine, M.A., Estabrooks, L.L., Severson, M.C., and Dobyns, W.B. 1995. Albright hereditary osteodystrophy and del(2) (q37.3) in four unrelated individuals. Am. J. Med. Genet. 58:1–7.

Phelan, M.C. 2008. Deletion 22q13.3 syndrome. Orphanet J. Rare Dis. 3:14.

Pinkel, D., Segraves, R., Sudar, D., Clark, S., Poole, I., Kowbel, D., Collins, C., Kuo, W.L., Chen, C., Zhai, Y., Dairkee, S.H., Ljung, B.M., Gray, J.W., and Albertson, D.G. 1998. High resolution analysis of DNA copy number variation using comparative genomic hybridization to microarrays. Nat. Genet. 20:207–211.

Poswillo, D. 1973. The pathogenesis of the first and second branchial arch syndrome. Oral Surg. Oral Med. Oral. Pathol. 35:302–328.

Potocki, L., Glaze, D., Tan, D.X., Park, S.S., Kashork, C.D., Shaffer, L.G., Reiter, R.J., and Lupski, J.R. 2000. Circadian rhythm abnormalities of melatonin in Smith-Magenis syndrome. J. Med. Genet. 37:428–433.

Prader, A., Labhart, A., and Willi, H. 1956. Ein Syndrom von Adipositas, Kleinwuchs, Kryptorchismus und Oligophrenie nach myatonieartigem Zustand im Neugeborenenalter. Schweiz. Med. Wochenschr. 86:1260–1261.

Pyeritz, R.E., and McKusick, V.A. 1979. The Marfan syndrome: diagnosis and management. N. Engl. J. Med. 300:772–777.

Richter, B., Schultealbert, A.H., and Koch, M.C. 2002. Human T and risk for neural tube defects. J. Med. Genet. 39:E14.

Rollnick, B.R., Kaye, C.I., Nagatoshi, K., Hauck, W., and Martin, A.O. 1987. Oculoauriculovertebral dysplasia and variants: phenotypic characteristics of 294 patients. Am. J. Med. Genet. 26:361–375.

Rose, P.S., Ahn, N.U., Levy, H.P., Ahn, U.M., Davis, J., Liberfarb, R.M., Nallamshetty, L., Sponseller, P.D., and Francomano, C.A. 2001. Thoracolumbar spinal abnormalities in Stickler syndrome. Spine 26:403–409.

Ross, A.J., Ruiz-Perez, V., Wang, Y., Hagan, D.M., Scherer, S., Lynch, S.A., Lindsay, S., Custard, E, Belloni, E., Wilson, D.I., Wadey, R., Goodman, F., Orstavik, K.H., Monclair, T., Robson, S., Reardon, W., Burn, J., Scambler, P., and Strachan, T. 1998. A homeobox gene, HLXB9, is the major locus for dominantly inherited sacral agenesis. Nat. Genet. 20:358–361.

Sáenz, A., Azpitarte, M., Armañanzas, R., Leturcq, F., Alzualde, A., Inza, I., García-Bragado, F., De la Herran, G., Corcuera, J., Cabello, A., Navarro, C., De la Torre, C., Gallardo, E., Illa, I., and de Munain, A.L. 2008. Gene expression profiling in limb-girdle muscular dystrophy 2A. PLoS ONE 3:e3750.

Samartzis, D.D., Herman, J., Lubicky, J.P., and Shen, F.H. 2006. Classification of congenitally fused cervical patterns in Klippel-Feil patients: epidemiology and role in the development of cervical spine-related symptoms. Spine 31:E798–804.

Schalkwijk, J., Zweers, M.C., Steijlen, P.M., Dean, W.B., Taylor, G., van Vlijmen, I.M., van Haren, B., Miller, W.L., and Bristow, J. 2001. A recessive form of the Ehlers-Danlos syndrome caused by tenascin-X deficiency. N. Engl. J. Med. 345:1167–1175.

Searle, A.G. 1966. Curtailed, a new dominant T-allele in the house mouse. Genet. Res. 7:86–95.

Shands, A.R. Jr., and Eisberg, H.B. 1955. The incidence of scoliosis in the state of Delaware: a study of 50,000 minifilms of the chest made during a survey for tuberculosis. J. Bone Joint Surg. Am. 37-A:1243–1249.

Shields, D.C., Ramsbottom, D., Donoghue, C., Pinjon, E., Kirke, P.N., Molloy, A.M., Edwards, Y.H., Mills, J.L., Mynett-Johnson, L., Weir, D.G., Scott, J.M., and Whitehead, A.S. 2000. Association between historically high frequencies of neural tube defects and the human T homologue of mouse T (Brachyury). Am. J. Med. Genet. 92:206–211.

Shores, J., Berger, K.R., Murphy, E.A., and Pyeritz, R.E. 1994. Progression of aortic dilatation and the benefit of long-term beta-adrenergic blockade in Marfan's syndrome. N. Engl. J. Med. 330:1335–1341.

Sköld, A.C., Wellfelt, K., and Danielsson, B.R. 2001. Stage-specific skeletal and visceral defects of the I(Kr)-blocker almokalant: further evidence for teratogenicity via a hypoxia-related mechanism. Teratology 64:292–300.

Slager, R.E., Newton, T.L., Vlangos, C.N., Finucane, B., and Elsea, S.H. 2003. Mutations in RAI1 associated with Smith-Magenis syndrome. Nat. Genet. 33:466–468.

Smith, A.C., Dykens, E., and Greenberg, F. 1998. Behavioral phenotype of Smith-Magenis syndrome (del 17p11.2). Am. J. Med. Genet. 81:179–185.

Smith, A.C., McGavran, L., Robinson, J., Waldstein, G., Macfarlane, J., Zonona, J., Reiss, J., Lahr, M., Allen, L., and Magenis, E. 1986. Interstitial deletion of (17)(p11.2p11.2) in nine patients. Am. J. Med. Genet. 24:393–414.

Snead, M.P., Yates, J.R., Pope, F.M., Temple, I.K., and Scott, J.D. 1996a. Masked confirmation of linkage between type 1 congenital vitreous anomaly and COL 2A1 in Stickler syndrome. Graefes. Arch. Clin. Exp. Ophthalmol. 234:720–721.

Snead, M.P., Yates, J.R., Williams, R., Payne, S.J., Pope, F.M., and Scott, J.D. 1996b. Stickler syndrome type 2 and linkage to the COL 11A1 gene. Ann. N. Y. Acad. Sci. 785: 331–332.

Speer, M.C., Melvin, E.C., Viles, K.D., Bauer, K.A., Rampersaud, E., Drake, C., George, T.M., Enterline, D.S., Mackey, J.F., Worley, G., Gilbert, J.R., and Nye, J.S. 2002. NTD Collaborative Group. Neural Tube Defects. T locus shows no evidence for linkage disequilibrium or mutation in American Caucasian neural tube defect families. Am. J. Med. Genet. 110: 215–218.

Steiner, C.E., Torriani, M., Norato, D.Y., and Marques-de-Faria, A.P. 2000. Spondylocarpotarsal synostosis with ocular findings. Am. J. Med. Genet. 91:131–134.

Steinmann, B., Royce, P.M., and Superti-Furga, A. 2002. The Ehlers-Danlos syndrome. In Connective Tissue and Its Heritable Disorders: Molecular, Genetic and Medical Aspects, 2nd Ed., eds. P.M. Royce and B. Steinmann, pp. 431–523. New York: Wiley-Liss

Stoll, C., Viville, B., Treisser, A., and Gasser, B. 1998. A family with dominant oculoauriculovertebral spectrum. Am. J. Med. Genet. 78:345–349.

Stossel, T.P., Condeelis, J., Cooley, L., Hartwig, J.H., Noegel, A., Schleicher, M., and Shapiro, S.S. 2001. Filamins as integrators of cell mechanics and signalling. Nat. Rev. Mol. Cell. Biol. 2:138–145.

Tassabehji, M., Fang, Z.M., Hilton, E.N., McGaughran, J., Zhao, Z., de Bock, C.E., Howard, E., Malass, M., Donnai, D., Diwan, A., Manson, F.D., Murrell, D., and Clarke, R.A. 2008. Mutations in GDF6 are associated with vertebral segmentation defects in Klippel-Feil syndrome. Hum. Mutat. 29:1017–1027.

Thauvin-Robinet, C., Faivre, L., Huet, F., Journeau, P., Glorion, C., Rustin, P., Rötig, A., Munnich, A., and Cormier-Daire, V. 2006. Another observation with VATER association and a complex IV respiratory chain deficiency. Eur. J. Med. Genet. 49:71–77.

Thomsen, B., Horn, P., Panitz, F., Bendixen, E., Petersen, A.H., Holm, L.E., Nielsen, V.H., Agerholm, J.S., Arnbjerg, J., and Bendixen, C. 2006. A missense mutation in the bovine SLC35A3 gene, encoding a UDP-N-acetylglucosamine transporter, causes complex vertebral malformation. Genome Res. 16:97–105.

Tian, Y., Ishikawa, H., Yamaguchi, T., Yamauchi, T., and Yokoyama, K. 2005. Teratogenicity and developmental toxicity of chlorpyrifos. Maternal exposure during organogenesis in mice. Reprod. Toxicol. 20:267–270.

Tracy, M.R., Dormans, J.P., and Kusumi, K. 2004. Klippel-Feil syndrome: clinical features and current understanding of etiology. Clin. Orthop. Relat. Res. 424:183–190.

Tredwell, S.J., Smith, D.F., Macleod, P.J., and Wood, B.J. 1982. Cervical spine anomalies in fetal alcohol syndrome. Spine 7:331–334.

Tribioli, C., Frasch, M., and Lufkin, T. 1997. Bapx1: an evolutionary conserved homologue of the Drosophila bagpipe homeobox gene is expressed in splanchnic mesoderm and the embryonic skeleton. Mech. Dev. 65:145–162.

Vissers, L.E., van Ravenswaaij, C.M., Admiraal, R., Hurst, J.A., de Vries, B.B., Janssen, I.M., van der Vliet, W.A., Huys, E.H., de Jong, P.J., Hamel, B.C., Schoenmakers, E.F., Brunner, H.G., Veltman, J.A., and van Kessel, A.G. 2004. Mutations in a new member of the chromodomain gene family cause CHARGE syndrome. Nat. Genet. 36:955–957.

Wang, X., Moreau, M., Raso, V.J., Zhao, J., Jiang, H., Mahood, J., and Bagnall, K. 1998. Changes in serum melatonin levels in response to pinealectomy in the chicken and its correlation with development of scoliosis. Spine 23:2377–2381.

Watabe-Rudolph, M., Schlautmann, N., Papaioannou, V.E., and Gossler, A. 2002. The mouse rib-vertebrae mutation is a hypomorphic Tbx6 allele. Mech. Dev. 119:251–256.

Weihing, R.R. 1985. The filamins: properties and functions. Can. J. Biochem. Cell. Biol. 63: 397–413.

Wéry, N., Narotsky, M.G., Pacico, N., Kavlock, R.J., Picard, J.J., and Gofflot, F. 2003. Defects in cervical vertebrae in boric acid-exposed rat embryos are associated with anterior shifts of hox gene expression domains. Birth Defects Res. A Clin. Mol. Teratol. 67:59–67.

Wynne-Davies, R. 1975. Congenital vertebral anomalies: aetiology and relationship to spina bifida cystica. J. Med. Genet. 12:280–288.

Xue, Y., Gao, X., Lindsell, C.E., Norton, C.R., Chang, B., Hicks, C., Gendron-Maguire, M., Rand, E.B., Weinmaster, G., and Gridley, T. 1999. Embryonic lethality and vascular defects in mice lacking the Notch ligand Jagged1. Hum. Mol. Genet. 8:723–730.

Yeowell, H.N., and Walker, L.C. 2000. Mutations in the lysyl hydroxylase 1 gene that result in enzyme deficiency and the clinical phenotype of Ehlers-Danlos syndrome type VI. Mol. Genet. Metab. 71:212–224.

Yoshiura, K.I., and Murray, J.C. 1997. Sequence and chromosomal assignment of human BAPX1, a bagpipe-related gene, to 4p16.1: a candidate gene for skeletal dysplasia. Genomics 45:425–428.

Zweers, M.C., Bristow, J., Steijlen, P.M., Dean, W.B., Hamel, B.C., Otero, M., Kucharekova, M., Boezeman, J.B., and Schalkwijk, J. 2003. Haploinsufficiency of TNXB is associated with hypermobility type of Ehlers-Danlos syndrome. Am. J. Hum. Genet. 73:214–217.

Chapter 8
Genetics and Functional Pathology of Idiopathic Scoliosis

Nancy H. Miller

Introduction

Idiopathic scoliosis is a fixed structural lateral curve of the spine of $\geq 10°$ affecting 2–3% of children making it the most common spinal abnormality in children. Clinically, it is characterized by a subtle, pain-free onset of bony deformity in otherwise normal individuals (see Fig. 8.1). Although the female-to-male ratio is approximately 1:1 for minor curves, the proportion of females increases dramatically as the curve magnitude increases for unknown reasons (Kane and Moe 1970). The wide variation of presentation, the limited therapeutic options, and the inability to detect individuals at risk for significant progression have led to the establishment of high-cost screening programs in the United States. These programs have resulted in high rates of subspecialty referrals and large numbers of radiographs for potential disease detection and analysis. With significant curve progression, spinal fusion may be the recommended treatment option. The economic cost attributable to scoliosis within the United States has been estimated to be as high as $4 billion/year and does not include the potential late morbidity in adulthood related to the long-term effects of a spinal fusion done at a young age. For these reasons, the study of idiopathic scoliosis has become a focus for further research in an effort to aid in the development of more effective targeted therapeutic options.

Classification and Natural History

Idiopathic scoliosis is classically divided into its age of clinical presentation, that is, infantile (0–3 years), juvenile (3–10 years), and adolescent (>10 years) (Kane and Moe 1970, Weinstein 1994). Due to the ambiguity surrounding the prepubertal stages of growth, this classification has been modified to "early onset" scoliosis

N.H. Miller (✉)
Professor, Department of Orthopaedic Surgery, Musculoskeletal Research Center, The Children's Hospital and University of Colorado Denver, Anschutz Medical Campus, Aurora, CO 80045, USA
e-mail: nancy.hadley-miller@ucdenver.edu

K. Kusumi, S.L. Dunwoodie (eds.), *The Genetics and Development of Scoliosis*,
DOI 10.1007/978-1-4419-1406-4_8, © Springer Science+Business Media, LLC 2010

Fig. 8.1 Clinical
presentation of an adolescent
female with idiopathic
scoliosis. Standing, one
notices the asymmetry of the
scapula and the prominence
of the right side of her back.
With forward bending, she is
noted to have a "rib hump"
due to trunk rotation

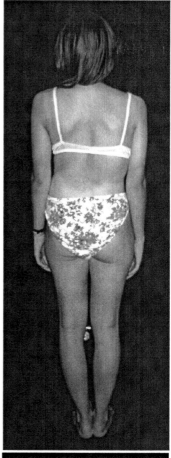

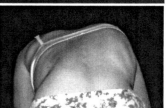

(presenting prior to the age of 5 years) and "later onset" scoliosis. Early onset scoliosis is frequently associated with underlying genetic or neurological disorders and has a variable history. Later onset scoliosis or "adolescent" idiopathic scoliosis, which is the subject of this discussion, occurs in the otherwise normal child with no underlying conditions and whose unknown etiology has frustrated clinicians for decades.

Our knowledge regarding the natural history of IS stems primarily from clinical studies of large population subgroups (Weinstein 1994, Bunnell 1988, Rogala et al. 1978). Once diagnosed through radiographic analysis, treatment decisions are based on clinical factors. Factors predicting curve progression include skeletal maturity, age at diagnosis, menarcheal status, curve size, curve pattern, and position of the curve apex (Weinstein 1994, Winter 1982). The more skeletally immature the patient, the greater the probability of curve progression. The more severe the curve at presentation, also, the higher the chance of curve progression. The most common curve pattern is a right thoracic curvature. Curves with a thoracic apex have the highest prevalence of progression, ranging from 58 to 100% in the immature individual (Lonstein and Carlson 1984, Weinstein and Ponseti 1983, Weinstein et al. 1981).

Once a diagnosis is made, the clinician utilizes the clinical prognostic factors to direct advice regarding conservative (nonoperative) or operative treatment (Weinstein 1994, Bunnell 1988, Rogala et al. 1978, Lonstein and Carlson 1984, Weinstein and Ponseti 1983, King et al. 1983, Kostuik 1990). Conservative treatment of back bracing is directed to the immature child with a significant chance of progression (Winter 1982, Lonstein and Carlson 1984, Winter et al. 1986). The goal of bracing is to stop the curve from progression; ultimately, it does not eliminate or improve the curvature. Back bracing (a thoracolumbar orthosis) imposes a significant psychosocial burden on the developing child (Clayson et al. 1987, Noonan et al. 1997). For severely progressive curves in both the immature and the mature individuals, surgical intervention is an option. Technically, instrumentation is used to achieve some curve correction, and then bone graft is placed to fuse the spine and prevent further progression of the curve. This procedure carries with it significant morbidity surrounding the surgical procedure itself, a loss of spinal mobility, and a limited knowledge of the longevity and long-term morbidity in any one patient (Winter 1982, King et al. 1983, Kostuik 1990, Szappanos et al. 1997). The clinical dilemma involves identifying the young individuals at highest risk for disease progression. A significant number of affected individuals undergo brace treatment due to known risk factors related to curve progression; however, in many cases, the treatment fails and surgical intervention is the only option to stop curve progression.

Overview of Genetic Analysis of Idiopathic Scoliosis

Although the specific cause of idiopathic scoliosis has not been determined, the role of hereditary or genetic factors in the development of this condition is widely accepted (Winter 1982, Beals 1973, Berquet 1966, Cowell et al. 1972, Dickson 1992, Harrington 1977, Willner 1994, Wynne-Davies 1968). Early clinical observations and population studies have documented a higher prevalence of scoliosis among relatives of affected individuals than in the general population (Wynne-Davies 1968, De George and Fisher 1967, Faber 1936, Garland 1934, Ma and Hu 1994, MacEwen and Shands 1967, Perricone and Paradiso 1987). The first formal population studies published in the English language literature by Harrington (1977), Wynne-Davies (1968), and Riseborough and Wynne-Davies (1973) document the increased incidence of scoliosis in families as compared with that in the general population. Harrington studied women whose scoliotic curves exceeded 15° and found an incidence of 27% of scoliosis among their daughters (Harrington 1977). Population studies involving index patients and their families indicate that 11% of first degree relatives are affected, as are 2.4 and 1.4% of second and third degree relatives, respectively (Wynne-Davies 1968, Riseborough and Wynne-Davies 1973). A recent study based on a unique population database from Utah (GenDB) concluded that within their population, 97% of idiopathic scoliosis patients have familial origins (Ogilvie et al. 2006). At least one major gene was determined with the suggestion of the presence of more than one gene within different families.

Twin studies have consistently shown monozygous twins to maintain a high concordance rate of approximately 73% for the condition. Dizygous twins may or may not be concordant in the expression of the disease, having a concordance rate of 36% (De George and Fisher 1967, Carr 1990, Gaertner 1979, Senda et al. 1997, Kesling and Reinker 1997, McKinley and Leatherman 1978, Murdoch 1959, Scott and Bailey 1963). This value is consistently higher than that reported of a first degree relative from more global population studies (Wynne-Davies 1968, Riseborough and Wynne-Davies 1973). This may be related to the high rate of radiographic confirmation of affectation status within twin studies as compared to that of large population studies of families. Radiographic confirmation of disease potentially lowers the false negative rate, as small scoliotic curves undetectable by clinical examination would then be ascertained as positive.

Despite these convincing arguments regarding the familial nature of scoliosis, the particular mode of heritability is unclear. Studies based on a wide variety of populations have suggested an autosomal dominant mode of inheritance, X-linked, or multifactorial inheritance pattern (Faber 1936, Garland 1934, Ballesteros et al. 1977, Bonaiti et al. 1976, Bell and Teebi 1995, Carr et al. 1992, Czeizel et al. 1978, Funatsu 1980, Miller et al. 1996, Mongird-Nakonieczna and Kozlowski 1976). The studies of Wynne-Davies and Riseborough of 2000 and 2869 individuals, respectively, suggested either a dominant or a multifactorial mode of inheritance (Wynne-Davies 1968, Riseborough and Wynne-Davies 1973). Cowell et al. (1972), noting relatively few male-to-male transmission in the literature, reported on 17 families as consistent with an X-linked dominant inheritance pattern. Given this controversy, Miller et al. investigated X-linkage as the mode of inheritance of familial idiopathic scoliosis in a subset of a large population of families with scoliosis (Justice et al. 2003, Miller et al. 1998). Results suggest that a region on the X (Xq23-26) chromosome may be linked to the expression of the disorder in a subset of families where the condition is segregating in an X-linked dominant manner.

Segregation Analysis

An alternative statistical method directed to the clarification of a genetic model and the penetrance of a familial disease is complex segregation analysis. This methodology can confirm clinical observations that a genetic determination exists for a specific disorder and can potentially aid in the delineation of an inheritance pattern. Aksenovich et al. (1988) used complex segregation analysis to study a population of 90 pedigrees with idiopathic scoliosis and their families (283 individuals). While their results confirmed the clinical observation of a genetic determination of this disorder in relationship to severe disease, milder forms of the disease were considered to be the result of heterogeneity. Axenovich et al. (1999) used complex segregation analysis to study 101 families (788 individuals). The results supported a genetic model; however, when the entire sample was considered, the best fitting genetic model was equivocal. When the authors excluded 27 families with probands with mild scoliosis (less than 25°), a Mendelian model with sex-dependant penetrance could not be rejected when compared with a general model, suggesting that the disorder was due to a single autosomal dominant locus ($p < 0.001$).

Heterogeneity and Other Confounding Factors

The variable results of the above studies emphasize the critical importance of diagnostic criteria and methods of disease ascertainment, which, in turn, are critical to the potential linkage of a specific disease state to genetic inheritability. Early studies of scoliosis are complicated by the fact that study populations may have included individuals with scoliosis associated with a primary disease condition, such as Marfan syndrome, rickets, or osteogenesis imperfecta (Berquet 1966, Faber 1936, Garland 1934, Shapiro et al. 1989). Later works frequently fail to detail the specific criteria for a positive affectation status, either clinically and/or radiographically. Finally, confirmation of affectation status and an ability to obtain a curve measurement through a spinal radiograph has not been critically adhered to in most reports (Wynne-Davies 1968, De George and Fisher 1967, Riseborough and Wynne-Davies 1973, Robin and Cohen 1975, Czeizel et al. 1978, Levaia 1981). Radiographic analysis is the definitive methodology for documentation of the true idiopathic scoliosis from one of alternative origins and for curve magnitude (Shands and Eisberg 1955). Multiple topographic methods have been utilized to study back asymmetry; however, there is no definitive correlation with underlying bony structure (Armstrong et al. 1982, Bunnell 1993). Careful history and clinical examination must eliminate features of alternative disorder in order to maintain the true idiopathic nature of the diagnosis (Shapiro et al. 1989, Levaia 1981, Boileau et al. 1993).

The above studies collectively attempt to indicate that scoliosis is a single gene disorder that follows the simple patterns of Mendelian genetics. This is dependent upon specific features characteristic to the study population. However, the simple patterns of Mendelian inheritance are known to be susceptible to the genetic principles of penetrance and heterogeneity. Penetrance is the probability that an individual carrying a gene will express the disorder. If penetrance is incomplete, a proportion of the individuals with the gene will not have the disorder. Genetic heterogeneity exists when there are different causes for the same disorder. Multiple genes with varying inheritance patterns could potentially result in the same clinical phenotype within or among families.

Positional and Candidate Gene Approaches

Prior to the use of genome-wide linkage analysis, the candidate gene approach was a classic approach directed to the identification of disease genes. Genes with known protein products that appear to be relevant to the physiological basis of the disease, such as connective tissue genes within the extracellular matrix, may be tested individually within patients, families, and/or identified sample populations (see Table 1).

Results of these studies have excluded genes for fibrillin 1 (*FBN1*) and 2 (*FBN2*); collagen type I (*COL1A1*) and II (*COL2A1*); elastin (*ELN*); aggrecan (*ACAN*); and heparin sulfotransferases (*HS3ST3A1* and *HS3ST3B1*) as causes of AIS (Miller et al. 1996, Marosy et al. 2006). Inoue et al. (2002a, b) studied polymorphisms in the genes for the vitamin D receptor (*MED4*), the estrogen receptor (*ESR1*), and cytochrome P450 family 17 (*CYP17A1*) in relation to curve progression. The results suggested that the XbaJ site polymorphism in *ESR1* was associated with curve progression. However, these results were refuted by Tang et al. (2006), who did not show an association between *ESR1* and adolescent idiopathic scoliosis. With the association of scoliosis with multiple connective tissues disorders, 30 multiple extracellular matrix genes have been reported to date, including the collagens, elastin, fibrillin 15, and aggrecan with no distinctive positive findings at the gene level (Carr 1990, Miller et al. 1996, Marosy et al. 2006, Zorkol'tseva et al. 2002).

Candidate genes have also been identified by linkage analysis and positional cloning efforts. Morcuende et al. (2003) focused on the human melatonin-1A receptor as a potential candidate gene for familial idiopathic scoliosis. Melatonin-1A receptor (hMel-1A) is located on chromosome 4q, a locus reported by Wise et al. to be potentially significant. Secondarily, experimental pinealectomy in chickens with resulting scoliosis has potentially related the melatonin pathway as a causative factor in the development of the condition (Machida et al. 1995, Bagnall et al. 1999). Through screening of the coding region of the HMel-1A receptor gene in 47 families (177 individuals), results showed no evidence for linkage of this gene and its variants to scoliosis in the study population (Morcuende et al. 2003).

Most recently, Moreau et al. (2004) have harvested osteoblasts from individuals with idiopathic scoliosis. Through culture techniques, responsiveness of the cells to reagents stimulatory to the melatonin pathway exhibits differences as compared to that of individuals without scoliosis (Moreau et al. 2004, Letellier et al. 2008). The potential relationship, significance, and clinical relevance of this finding are still unknown.

Merola et al. (2003) reported a potential association of a polymorphism within the aggrecan gene on chromosome 15q with idiopathic scoliosis. Marosy et al. (2006) recognized a specific subgroup of patients (48 families) that showed significant results from a genome-wide screen on chromosome 15q25–26 where this gene lies. In an effort to replicate the report's finding, the investigators conducted a study to determine the potential association between the identified aggrecan gene polymorphism and the familial scoliosis. Results showed no evidence of association between the polymorphism and scoliosis or the degree of lateral curvature.

A recent study by Gao et al. (2007) pursued studies on the chromodomain helicase DNA-binding protein 7 (*CHD7*) gene within a previously identified region on chromosome 8 by their own laboratory. This is discussed further in the following chapter.

Chromosomal Anomalies Associated with Idiopathic Scoliosis

Another strategy for the identification of disease genes is the use of an identified chromosomal alteration in affected individuals. Based on a report of a potential co-segregation of a balanced chromosomal rearrangement with scoliosis on a small pedigree, Bashiardes et al. (2004) pursued this area for a potential causative gene. They demonstrated that the q-arm break occurs in the gamma-1-syntrophin (*SNTG1*) gene. Syntrophins are proteins of the cytoplasmic membrane that are believed to associate with dystrophin, the product responsible for Duchenne muscular dystrophy, making them a reasonable candidate for the potential development of scoliosis. The authors went further to look at the gene in additional families. Despite the finding of two genetic variants, the change in *SNTG1* did not co-segregate with the disease in all affected individuals.

Genome-Wide Linkage Analysis

Advances in the mapping of the human genome and genetic technology now allow screening of the entire genome of an individual with known genetic markers or polymorphisms evenly spaced along the chromosomes. Genetic polymorphisms are benign variations within the DNA structure that vary from individual to individual. Their potential physical proximity to a nucleotide sequence that is a functional aspect of the genome allows localization of specific genomic area(s) and determination of their association with an expressed condition. This type of analysis can be performed on individuals or study populations. The current literature now includes multiple studies which have utilized statistical linkage analysis to identify specific genetic loci related to the expression of familial idiopathic scoliosis (Justice et al. 2003, Alden et al. 2006, Chan et al. 2002, Miller et al. 2005, Salehi et al. 2002, Wise et al. 2000). Each study has been constructed around a different set of variables with the application of a variety of genetic statistical programs; therefore, despite their differing results, they should not be construed as conflicting in nature.

Wise et al. (2000) reported a genome scan of families with more severe scoliosis segregating in an autosomal dominant pattern. One extended family (seven affected individuals) showed suggestive evidence for linkage of the disease state on chromosomes 6, 10, 12, and 18. When these areas were studied in four additional families, one family showed supportive evidence again for genetic linkage on chromosome 10.

Two studies (Chan et al. 2002, Salehi et al. 2002) again illustrate the application of statistical linkage analysis within two separate study populations. The first, by

Table 8.1 A review of studies related to the genetic etiology of idiopathic scoliosis

Study	Approach[1]/ Type of sample	Region	Candidate gene(s)	Boundaries	p-value
Carr et al. 1992	Linkage/4 families	17q21; 7q22	COLIA1, COLIA2, COL2A1	NA	NS
Miller et al. 1996	Linkage/11 families	15q21.1; 7q11; 7q22	ELN, FBN1, COLIA2	NA	NS
Zorkol'tseva et al. 2002	Association/33 families	15	AGC1 (exon G3)	NA	NS
Morcuende et al. 2003	Linkage/47 families	4q35	MTNR1A	NA	NS
Marosy et al. 2006	Linkage, association/58 families	15	AGC1 (exon 12)	NA	NS
Montanaro et al. 2006	Association/81 trios	1p35	MATN1	30.96–30.97	0.024
Qiu et al. 2006	Association/20 cases	11q21	MTNR1B	NA	NS
Tang et al. 2006	Association/540 cases	6q25	ESR1	NA	NS
Wu et al. 2006	Association/202 cases	6q25	ESR1	NA	0.001
Yeung et al. 2006	Association/506 cases	12q22	IGF1*	NA	NS
Wise et al. 2000	Linkage/2 families	6q	NA	NA	0.023[2]
		Distal 10q	NA	NA	0.0193[2]
		18q	NA	NA	0.0023[2]
Chan et al. 2002	Linkage/7 families	19p13.3	D19S894–D19S1034	4.34–6.06	0.00001 (4.48)[3]
		2q13–2q22.3	D2S160–D2S151	113.09–148.00	0.0049 (1.72)[3]
Salehi et al. 2002	Linkage/1 family	17p11–17q11.2	D17S799–D17S925	13.37–27.45	0.0001 (3.2)[3]
Justice et al. 2003	Linkage/51 families	Xq23–Xq26.1	DXS6804–DXS1047	110.88–127.78	0.0014 (2.23)[3]
		6p	F13A1–D6S2439	6.09–24.42	0.01215[4]
		6q16	D6S1031–D6S1021	77.46–104.72	0.00215[4]
		9q32–9q34	D9S938–D9S1838	101.36–135.86	0.00055[4]
		16q11–16q12	D16S764–D16S3253	16.61–54.57	0.00025[4]
Miller et al. 2005	Linkage/202 families	17p11–17q11	D17S1303–D17S1293	11.06–32.71	0.0025[4]
Alden et al. 2006	Linkage/72 families	19p13	D19S591–D19S1034	3.01–6.05	0.013565[4]

Table 8.1 (continued)

Study	Approach[1] / Type of sample	Region	Candidate gene(s)	Boundaries	p-value
Miller et al. 2006	Linkage, association/7 families	5q13	D5S417–D5S807	3.17–9.26	0.00173[2]
		13q13..3	D13S305–D13S788	35.92–50.79	0.00013[2]
		13q32	D13S800–D13S779	72.77–100.3	0.00013[2]
Gao et al. 2007	Linkage, association/52 families	8q12	CHD7	61.75–61.94	0.005
Ocaka et al. 2008	Linkage/25 families	9q31.2–q34.2	D9S930–D9S1818	114.18–136.18	0.00004 (3.64)[3]
		17q25.3–qtel	D17S1806	74.85–qtel	0.00001 (4.08)[3]

[1] Linkage or association study (familial or case control)
[2] Most significant p-value obtained, originating from multipoint linkage analysis
[3] Most significant p-value obtained, calculated from a Lod score
[4] Most significant p-value obtained, originating from single-point linkage analysis
* Insulin-like growth factor 1
NA = not applicable; NS = not significant
Map positions from the UCSC Human Genome Working Draft website (http://www.genome.ucsc.edu/) build 33

Chan et al. (2002), involved seven multiplex families identified through a proband with a minimum Cobb angle of 20°. The authors first screened a multiplex family for potential linkage on chromosomes 6p, 10q, and 18q as reported by Wise et al. (2000). Failure to confirm linkage led to a genome-wide scan of this family. Results indicated a primary area on chromosome 19p13.3 and a secondary area on chromosome 2q of potential importance. Six additional kindreds were tested, lending further support to the initial finding. A second work by Salehi et al. (2002) investigated a single three-generation Italian family of 17 individuals. Eleven individuals were deemed affected, with radiographs exhibiting a spinal curvature of at least 10°. Analyses indicated an area on chromosome 17p11 of potential importance. The authors recognized candidate genes within this area because of their role in extracellular matrix structure and integrity. Sequencing of two of the genes in two affected individuals and a control subject failed to identify any pathogenic sequence variations.

Miller et al. (2005) reported on a genomic screen of a large sample of families (202 families, 1198 individuals). Families were determined through a proband of at least a 10° lateral curvature. The large sample population was stratified according to potential mode of inheritance, autosomal dominant versus X-linked dominant in nature. Within the sample group deemed to represent an autosomal dominant mode of heritability, analyses resulted in five primary (chromosomes 6p, 6q, 9, 16, and 17) and eight secondary (chromosomes 1, 3, 5, 7, 8, 11, 12, 19) potential areas of significance. Fine-mapping efforts have both corroborated and refined many of these areas. The authors have continued to stratify the large population based on clinical criteria such as left thoracic curves, kyphoscoliosis,, and curve severity. This has resulted in the identification of areas on chromosomes 5 and 13 as related to the expression of kyphoscoliosis (Justice et al. 2003, Miller et al. 2006). In a subset of families with more severe curves (>40°), an area on chromosome 19 has proven significant (Alden et al. 2006).

Summary

In conclusion, familial idiopathic scoliosis is a complex orthopedic disorder with multiple potential genes involved in its expression. The search for the underlying genetic mechanisms is heavily influenced by clinical and genetic heterogeneity. Successful associations of specific families and subgroups with specific genetic loci are an important step in the identification of the gene(s) related to this disorder. A better understanding of the genetic influences on this condition and resulting clinical variability will potentially improve the ability to assign a more specific prognosis and to develop more selective treatment regimens. This knowledge will also afford us deeper insight into the genetic influences related to skeletal development and spinal stability.

Acknowledgments We would like to thank Kandice Swindle, Shane Cook, Jeffrey Dunn for editorial assistance.

References

Aksenovich, T.I., Semenov, I.R., Ginzburg, E., and Zaidman, A.M. 1988. Preliminary analysis of inheritance of scoliosis. Genetika 24(11):2056–2063.

Alden, K.J., Marosy, B., Nzegwu, N., Justice, C.M., Wilson, A.F., and Miller, N.H. 2006. Idiopathic scoliosis: identification of candidate regions on chromosome 19p13. Spine 31(16):1815–1819.

Armstrong, G.W., Livermore, N.B., 3rd, Suzuki, N., and Armstrong, J.G. 1982. Nonstandard vertebral rotation in scoliosis screening patients: its prevalence and relation to the clinical deformity. Spine 7(1):50–54.

Axenovich, T.I., Zaidman, A.M., Zorkoltseva, I.V., Tregubova, I.L., and Borodin, P.M. 1999. Segregation analysis of idiopathic scoliosis: demonstration of a major gene effect. Am. J. Med. Genet. 86(4):389–394.

Bagnall, K., Raso, V.J., Moreau, M., Mahood, J., Wang, X., and Zhao, J. 1999. The effects of melatonin therapy on the development of scoliosis after pinealectomy in the chicken. J. Bone Joint Surg. Am. 81(2):191–199.

Ballesteros, S., Grove, H.M., Campusano, C., Teuber, A.M., and Figueroa, H. 1977. Genetic study in a family affected of idiopathic scoliosis (author's transl). Rev. Med. Chil. 105(4):224–226.

Bashiardes, S., Veile, R., Allen, M., Wise, C.A., Dobbs, M., Morcuende, J.A., Szappanos, L., Herring, J.A., Bowcock, A.M., and Lovett, M. 2004. SNTG1, the gene encoding gamma1-syntrophin: a candidate gene for idiopathic scoliosis. Hum. Genet. 115(1):81–89.

Beals, R.K. 1973. Nosologic and genetic aspects of scoliosis. Clin. Orthop. Relat. Res. 93:23–32.

Bell, M. and Teebi, A.S. 1995. Autosomal dominant idiopathic scoliosis? Am. J. Med. Genet. 55(1):112.

Berquet, K.H. 1966. Considerations on heredity in idiopathic scoliosis. Z. Orthop. Ihre Grenzgeb. 101(2):197–209.

Boileau, C., Jondeau, G., Babron, M.C., Coulon, M., Alexandre, J.A., Sakai, L., Melki, J., Delorme, G., Dubourg, O., Bonaiti-Pellie, C. et al. 1993. Autosomal dominant Marfan-like connective-tissue disorder with aortic dilation and skeletal anomalies not linked to the fibrillin genes. Am. J. Hum. Genet. 53(1):46–54.

Bonaiti, C., Feingold, J., Briard, M.L., Lapeyre, F., Rigault, P., and Guivarch, J. 1976. Genetics of idiopathic scoliosis. Helv. Paediatr. Acta 31(3):229–240.

Bunnell, W.P. 1988. The natural history of idiopathic scoliosis. Clin. Orthop. Relat. Res. 229:20–25.

Bunnell, W.P. 1993. Outcome of spinal screening. Spine 18(12):1572–1580.

Carr, A.J. 1990. Adolescent idiopathic scoliosis in identical twins. J. Bone Joint Surg. Br. 72(6):1077.

Carr, A.J., Ogilvie, D.J., Wordsworth, B.P., Priestly, L.M., Smith, R., and Sykes, B. 1992. Segregation of structural collagen genes in adolescent idiopathic scoliosis. Clin. Orthop. Relat. Res. 274:305–310.

Chan, V., Fong, G.C., Luk, K.D., Yip, B., Lee, M.K., Wong, M.S., Lu, D.D., and Chan, T.K. 2002. A genetic locus for adolescent idiopathic scoliosis linked to chromosome 19p13.3. Am. J. Hum. Genet. 71(2):401–406.

Clayson, D., Luz-Alterman, S., Cataletto, M.M., and Levine, D.B. 1987. Long-term psychological sequelae of surgically versus nonsurgically treated scoliosis. Spine 12(10):983–986.

Cowell, H.R., Hall, J.N., and MacEwen, G.D. 1972. Genetic aspects of idiopathic scoliosis. A Nicholas Andry Award essay, 1970. Clin. Orthop. Relat. Res. 86:121–131.

Czeizel, A., Bellyei, A., Barta, O., Magda, T., and Molnar, L. 1978. Genetics of adolescent idiopathic scoliosis. J. Med. Genet. 15(6):424–427.

De George, F.V. and Fisher, R.L. 1967. Idiopathic scoliosis: genetic and environmental aspects. J. Med. Genet. 4(4):251–257.

Dickson, R.A. 1992. The etiology and pathogenesis of idiopathic scoliosis. Acta Orthop. Belg. 58(Suppl 1):21–25.

Faber, A. 1936. Untersuchungen uber die Erblichkeit der Skoliose. Arch. Orthop. Unfallchir. 36:247–249.

Funatsu, K. 1980. Familial incidence in idiopathic scoliosis (author's transl). Nippon Seikeigeka Gakkai Zasshi. 54(7):633–649.

Gaertner, R.L. 1979. Idiopathic scoliosis in identical (monozygotic) twins. South. Med. J. 72(2):231–234.

Gao, X., Gordon, D., Zhang, D., Browne, R., Helms, C., Gillum, J., Weber, S., Devroy, S., Swaney, S., Dobbs, M., Morcuende, J., Sheffield, V., Lovett, M., Bowcock, A., Herring, J., and Wise, C. 2007. CHD7 gene polymorphisms are associated with susceptibility to idiopathic scoliosis. Am. J. Hum. Genet. 80(5):957–965.

Garland, H.G. 1934. Hereditary scoliosis. Br. Med. J. 1:328.

Harrington, P.R. 1977. The etiology of idiopathic scoliosis. Clin. Orthop. Relat. Res.126:17–25.

Inoue, M., Minami, S., Nakata, Y., Kitahara, H., Otsuka, Y., Isobe, K., Takaso, M., Tokunaga, M., Nishikawa, S., Maruta, T., and Moriya, H. 2002a. Association between estrogen receptor gene polymorphisms and curve severity of idiopathic scoliosis. Spine 27(21):2357–2362.

Inoue, M., Minami, S., Nakata, Y., Takaso, M., Otsuka, Y., Kitahara, H., Isobe, K., Kotani, T., Maruta, T., and Moriya, H. 2002b. Prediction of curve progression in idiopathic scoliosis from gene polymorphic analysis. Stud. Health Technol. Infor. 91:90–96.

Justice, C.M., Miller, N.H., Marosy, B., Zhang, J., and Wilson, A.F. 2003. Familial idiopathic scoliosis: evidence of an X-linked susceptibility locus. Spine 28(6):589–594.

Kane, W.J., and Moe, J.H. 1970. A scoliosis-prevalence survey in Minnesota. Clin. Orthop. Relat. Res. 69:216–218.

Kesling, K.L. and Reinker, K.A. 1997. Scoliosis in twins. A meta-analysis of the literature and report of six cases. Spine 22(17):2009–2014.

King, H.A., Moe, H., Bradford, D.S., and Winter, R.B. 1983. The selection of fusion levels in thoracic idiopathic scoliosis. J. Bone Joint Surg. Am. 65(9):1302–1313.

Kostuik, J.P. 1990. Operative treatment of idiopathic scoliosis. J. Bone Joint Surg. Am. 72(7): 1108–1013.

Letellier, K., Azeddine, B., Parent, S., Labelle, H., Rompre, P.H., Moreau, A., and Moldovan, F. 2008. Estrogen cross-talk with the melatonin signaling pathway in human osteoblasts derived from adolescent idiopathic scoliosis patients. J. Pineal Res. 45(4):383–393.

Levaia, N.V. 1981. Genetic aspect of dysplastic (idiopathic) scoliosis. Ortop. Travmatol. Protez. 2:23–29.

Lonstein, J.E., and Carlson, J.M. 1984. The prediction of curve progression in untreated idiopathic scoliosis during growth. J. Bone Joint Surg. Am. 66(7):1061–1071.

Ma, X.J. and Hu, P. 1994. The etiological study of idiopathic scoliosis. Zhonghua Wai Ke Za Zhi. 32(8):504–506.

MacEwen, G.D., and Shands, A.R., Jr. 1967. Scoliosis–a deforming childhood problem. Clin. Pediatr. (Phila) 6(4):210–216.

Machida, M., Dubousset, J., Imamura, Y., Iwaya, T., Yamada, T., and Kimura, J. 1995. Role of melatonin deficiency in the development of scoliosis in pinealectomised chickens. J. Bone Joint Surg. Br. 77(1):134–138.

Marosy, B., Justice, C.M., Nzegwu, N., Kumar, G., Wilson, A.F., and Miller, N.H. 2006. Lack of association between the aggrecan gene and familial idiopathic scoliosis. Spine 31(13): 1420–1425.

McKinley, L.M. and Leatherman, K.D. 1978. Idiopathic and congenital scoliosis in twins. Spine. 3(3):227–229.

Merola, A., Mathur, S., Igobou, S., Brkaria, M., Vigna, F., Paulino, C., Kubec, J., Haher, T., and Espat, N. 2003. Polymorphism of the aggrecan gene as a marker for adolescent idiopathic scoliosis. SRS 38th Annual Meeting, 2003, Quebec.

Miller, N.H., Mims, B., Child, A., Milewicz, D.M., Sponseller, P., and Blanton, S.H. 1996. Genetic analysis of structural elastic fiber and collagen genes in familial adolescent idiopathic scoliosis. J. Orthop. Res. 14(6):994–999.

Miller, N.H., Schwab, D.L., Sponseller, P., Shugert, E., Bell, J., and Maestri, N. 1998. Genomic search for X-linkage in familial adolescent idiopathic scoliosis. In Research into Spinal Deformities 2, ed. IRSoSDMn, pp. 209–213. Amsterdam: IOS Press

Miller, N.H., Justice, C.M., Marosy, B., Doheny, K.F., Pugh, E., Zhang, J., and Dietz, H.C., 3rd, Wilson, A.F. 2005. Identification of candidate regions for familial idiopathic scoliosis. Spine 30(10):1181–1187.

Miller, N.H., Marosy, B., Justice, C.M., Novak, S.M., Tang, E.Y., Boyce, P., Pettengil, J., Doheny, K.F., Pugh, E.W., and Wilson, A.F. 2006. Linkage analysis of genetic loci for kyphoscoliosis on chromosomes 5p13, 13q13.3, and 13q32. Am. J. Med. Genet. A 140(10):1059–1068.

Mongird-Nakonieczna, J. and Kozlowski, B. 1976. Familial occurrence of idiopathic scoliosis. Chir. Narzadow Ruchu Ortop. Pol. 41(2):161–165.

Montanaro, L., Parisini, P., Greggi, T., Di Silvestre, M., Campoccia, D., Rizzi, S., and Arciola, C.R. 2006. Evidence of a linkage between matrilin-1 gene (MATN1) and idiopathic scoliosis. Scoliosis 1:21.

Morcuende, J.A., Minhas, R., Dolan, L., Stevens, J., Beck, J., Wang, K., Weinstein, S.L., and Sheffield, V. 2003. Allelic variants of human melatonin 1A receptor in patients with familial adolescent idiopathic scoliosis. Spine 28(17):2025–2028.

Moreau, A., Wang, D.S., Forget, S., Azeddine, B., Angeloni, D., Fraschini, F., Labelle, H., Poitras, B., Rivard, C.H., and Grimard, G. 2004. Melatonin signaling dysfunction in adolescent idiopathic scoliosis. Spine 29(16):1772–1781.

Murdoch, G. 1959. Scoliosis in twins. J. Bone Joint Surg. Br. 41-B:736–737.

Noonan, K.J., Dolan, L.A., Jacobson, W.C., and Weinstein, S.L. 1997. Long-term psychosocial characteristics of patients treated for idiopathic scoliosis. J. Pediatr. Orthop. 17(6):712–717.

Ogilvie, J.W., Braun, J., Argyle, V., Nelson, L., Meade, M., and Ward, K. 2006. The search for idiopathic scoliosis genes. Spine 31(6):679–681.

Ocaka, L., Zhao, C., Reed, J.A., Ebenezer, N.D., Brice, G., Morley, T., Mehta, M., O'Dowd, J., Weber, J.L., Hardcastle, A.J., and Child, A.H. 2008. Assignment of two loci for autosomal dominant adolescent idiopathic scoliosis to chromosomes 9q31.2-q34.2 and 17q25.3-qtel. J. Med. Genet. 45(2):87–92.

Perricone, G. and Paradiso, T. 1987. Familial factors in so-called idiopathic scoliosis. Chir. Organi. Mov. 72(4):355–358.

Qiu, X.S., Tang, N.L., Yeung, H.Y., Qiu, Y., Qin, L., Lee, K.M., and Cheng, J.C. 2006. The role of melatonin receptor 1B gene (MTNR1B) in adolescent idiopathic scoliosis–a genetic association study. Stud. Health Technol. Inform. 123:3–8.

Riseborough, E.J. and Wynne-Davies, R. 1973. A genetic survey of idiopathic scoliosis in Boston, Massachusetts. J. Bone Joint Surg. Am. 55(5):974–982.

Robin, G.C. and Cohen, T. 1975. Familial scoliosis. A clinical report. J. Bone Joint Surg. Br. 57(2):146–148.

Rogala, E.J., Drummond, D.S., and Gurr, J. 1978. Scoliosis: incidence and natural history. A prospective epidemiological study. J. Bone Joint Surg. Am. 60(2):173–176.

Salehi, L.B., Mangino, M., De Serio, S., De Cicco, D., Capon, F., Semprini, S., Pizzuti, A., Novelli, G., and Dallapiccola, B. 2002. Assignment of a locus for autosomal dominant idiopathic scoliosis (IS) to human chromosome 17p11. Hum. Genet. 111(4–5):401–404.

Scott, T.F. and Bailey, R.W. 1963. Idiopathic scoliosis in fraternal twins. J. Mich. State Med. Soc. 62:283–284.

Senda, M., Harada, Y., Nakahara, S., and Inoue, H. 1997. Lumbar spinal changes over 20 years after posterior fusion for idiopathic scoliosis. Acta Med. Okayama 51(6):327–331.

Shands, A.R., Jr. and Eisberg, H.B. 1955. The incidence of scoliosis in the state of Delaware; a study of 50,000 minifilms of the chest made during a survey for tuberculosis. J. Bone Joint Surg. Am. 37-A(6):1243–1249.

Shapiro, J.R., Burn, V.E., Chipman, S.D., Velis, K.P., and Bansal, M. 1989. Osteoporosis and familial idiopathic scoliosis: association with an abnormal alpha 2(I) collagen. Connect. Tissue Res. 21(1–4):117–123.

Szappanos, L., Balogh, E., Szeszak, F., Olah, E., Nagy, Z., and Szepesi, K. 1997. Idiopathic scoliosis–new surgical methods or search for the reasons. Acta Chir. Hung. 36(1–4):343–345.

Tang, N.L., Yeung, H.Y., Lee, K.M., Hung, V.W., Cheung, C.S., Ng, B.K., Kwok, R., Guo, X., Qin, L., and Cheng, J.C. 2006. A relook into the association of the estrogen receptor [alpha] gene (PvuII, XbaI) and adolescent idiopathic scoliosis: a study of 540 Chinese cases. Spine 31(21):2463–2468.

Weinstein, S.L. 1994. Adolescent idiopathic scoliosis: prevalence and natural history. In The Pediatric Spine: Principles and Practice, ed. S.L. Weinstein, pp. 463–478. New York: Raven Press

Weinstein, S.L., and Ponseti, I.V. 1983. Curve progression in idiopathic scoliosis. J. Bone Joint Surg. Am. 65(4):447–455.

Weinstein, S.L., Zavala, D.C, and Ponseti, I.V. 1981. Idiopathic scoliosis: long-term follow-up and prognosis in untreated patients. J. Bone Joint Surg. Am. 63(5):702–712.

Willner, S. 1994. Continuous screening and treatment of teenage scoliosis is recommended. Lakartidningen 91(1–2):22.

Winter, R.B. 1982. Evolution in the treatment of idiopathic scoliosis in Minnesota. A family report. Minn. Med. 65(10):627–629.

Winter, R.B., Lonstein, J.E., Drogt, J., and Noren, C.A. 1986. The effectiveness of bracing in the nonoperative treatment of idiopathic scoliosis. Spine 11(8):790–791.

Wise, C.A., Barnes, R., Gillum, J., Herring, J.A., Bowcock, A.M., and Lovett, M. 2000. Localization of susceptibility to familial idiopathic scoliosis. Spine 25(18):2372–2380.

Wu, J., Qiu, Y., Zhang, L., Sun, Q., Qiu, X., and He, Y. 2006. Association of estrogen receptor gene polymorphisms with susceptibility to adolescent idiopathic scoliosis. Spine 31(10):1131–1136.

Wynne-Davies, R. 1968. Familial (idiopathic) scoliosis. A family survey. J. Bone Joint Surg. Br. 50(1):24–30.

Yeung, H.Y., Tang, N.L., Lee, K.M., Ng, B.K., Hung, V.W., Kwok, R., Guo, X., Qin, L., and Cheng, J.C. 2006. Genetic association study of insulin-like growth factor-I (IGF-I) gene with curve severity and osteopenia in adolescent idiopathic scoliosis. Stud. Health Technol. Inform. 123:18–24.

Zorkol'tseva, I.V., Liubinskii, O.A., Sharipov, R.N., Zaidman, A.M., Aksenovich, T.I., and Dymshits, G.M. 2002. Analysis of polymorphism of the number of tandem repeats in the aggrecan gene exon G3 in the families with idiopathic scoliosis. Genetika 38(2):259–263.

Chapter 9
Current Understanding of Genetic Factors in Idiopathic Scoliosis

Carol A. Wise and Swarkar Sharma

Introduction

Scoliosis

"Scoliosis" is derived from the Greek word meaning "crooked" and was used for the first time by Galen (AD 131–201) to describe an "S-shaped" or "C-shaped" spinal deformity (Fig. 9.1). Although defined as a lateral curvature as visualized by plane radiography, the deformity is actually three-dimensional and involves changes in the frontal, sagittal, and transverse planes of the spinal column. Patients treated for scoliosis generally belong to one of the three categories. In so-called congenital scoliosis, the structural curvature of the spine is clearly secondary to radiographically visible vertebral malformations and is typically obvious at an early age. Other patients may have scoliosis as part of other pathological conditions. For example, patients with neurologic or neuromuscular diseases such as Duchenne muscular dystrophy, spinal muscular atrophy, neurofibromatosis, or Charcot–Marie–Tooth disease may develop scoliosis possibly due to secondary weakness of the paravertebral muscles (Wise et al. 2008). Patients with other syndromes such as Prader–Willi (Yamada et al. 2007) or CHARGE (Doyle and Blake 2005) may develop an adolescent-onset scoliosis for reasons that are not as yet understood. However, the great majority (>80%) of scoliosis patients are otherwise healthy with no obvious coexisting diagnoses or structural alterations of the spinal column. This third class of patients is considered "idiopathic." Idiopathic scoliosis, which we will refer to as "IS," is the subject of the remainder of this chapter.

Idiopathic Scoliosis

IS is the most frequently treated form of scoliosis as noted, and also represents the most common spinal deformity in children. Clinically it is often described by

C.A. Wise (✉)
Texas Scottish Rite Hospital for Children, Dallas, TX 75219, USA
e-mail: carol.wise@tsrh.org

K. Kusumi, S.L. Dunwoodie (eds.), *The Genetics and Development of Scoliosis*,
DOI 10.1007/978-1-4419-1406-4_9, © Springer Science+Business Media, LLC 2010

Fig. 9.1 Spinal curvature in scoliosis

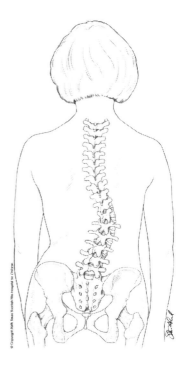

its age of onset as "infantile" (0–3 years), "juvenile" (3–9 years), or "adolescent" (10–18 years) (James 1954). However, if considered in broad developmental terms the onset of IS is highly correlated with periods of rapid vertical growth at two stages, infancy and preadolescence/adolescence. The later form of IS occurring with the adolescent "growth spurt" accounts for the majority of cases. Although estimates vary, adolescent IS generally affects 2–3% of school age children and is apparent in all ethnic population groups (see more on this below) (Herring 2002). Thus IS is a global health-care problem mostly affecting adolescent children and adults.

Presentation, Disease Course, and Treatment

IS can be a rapidly progressive disease, and therefore early detection is warranted. This has prompted many communities to institute screening programs in the schools for children approaching adolescence (Richards and Vitale 2008). Initial indicators are typically shoulder asymmetry and/or a noticeable "rib hump" on physical examination (Fig. 9.2). A diagnosis of IS requires further evaluation, including postero-anterior radiographs to quantify the degree of spinal curvature, and to exclude coexisting diagnoses. Although the spinal deformity in IS is actually three-dimensional as noted, it is typically quantified in two dimensions using the "Cobb angle method." The Cobb angle is a measure of the lateral deviation of the spine

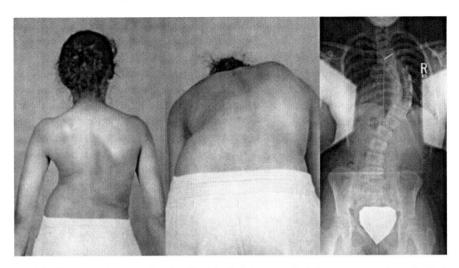

Fig. 9.2 IS in an adolescent female. Shoulder imbalance and rib hump are noted on physical examination in the *left* and *middle* panels. A standing postero-anterior radiograph reveals a right thoracic curve of the spinal column measuring 60° by the Cobb angle method and with no visible vertebral anomalies

from the vertical, where positive diagnosis of scoliosis is defined as deviation of more than 10° (Fig. 9.2). Disease progression is then defined as an increase in the Cobb angle measurement over time, and it is largely this measure together with other clinical markers that triggers treatment decisions as outlined below.

Although disfiguring, the majority of IS cases are stable (i.e., nonprogressive) and do not require intervention. However, these two outcomes are difficult if not impossible to distinguish in an initial examination. Identifying better predictive markers for this purpose is a major goal of IS research. Certain developmental and clinical markers are somewhat reliable in this regard and are typically monitored in present-day practice (Herring 2002). The most obvious of these is gender. For reasons that are not clear, girls are at least five times more likely to have progressive curves as compared to boys (Karol et al. 1993). Other risk factors are initial curve pattern, apical vertebral rotation, age at onset, and severity of curvature in relation to remaining growth. Typical measures to determine remaining growth in the child first presenting with IS include menarchal status in girls and bone age. Bone age is usually estimated from ossification of the iliac apophysis and/or tri-radiate cartilages of the pelvis that are visualized from the same radiograph used to measure the Cobb angle deformity (Herring 2002). Certain curve patterns, e.g., in the thoracic region, are also somewhat predictive of progression [reviewed in (Wise et al. 2008)]. Thus, although the various clinical markers may be suggestive of progressive (and serious) disease, none are sufficient to exclude the possibility, and most affected children are frequently monitored by radiographs and physical examination to detect progressive deformity. In the meantime, it is well-established that risk of curve progression falls off substantially after the child has reached peak growth velocity. Hence growth measurements are typically collected at each clinic visit and compared to historical peak growth-velocity data to determine if this important developmental milestone has been reached (Herring 2002). If curve progression is minimal or controlled through the high-risk period preceding this, e.g., by bracing, then surgery may be avoided. However, surgery is necessary for curves that continue to progress (generally to a threshold of ~50° Cobb angle) before the high-risk period is over (Herring 2002). If progressive scoliosis is neglected it can cause significant spinal deformity and alterations of the thoracic cage that ultimately evoke cardiopulmonary compromise (Pehrsson et al. 1991). Otherwise the natural history of untreated IS involves disfigurement, increased back pain, and spinal osteoarthritis (Herring 2002).

What Is Expected from This Chapter?

Despite many clinical investigations of IS, etiological understanding remains poor (reviewed in Lowe et al. 2000). As discussed quantitatively in this chapter, the genetic underpinnings in IS are significant and were established by literature from the past 70 years. Here we describe IS as a complex genetic disease and compare it to other common diseases in terms of genetic risk. We review recent linkage and association studies that likely explain only a small fraction of this genetic risk. Candidate susceptibility and modifier genes are also discussed in the context of these recent discoveries. Given the success of genome-wide approaches in identifying genetic loci underlying common diseases such as inflammatory bowel disease, psoriasis, age-related macular degeneration, type 2 diabetes, and coronary heart disease (Ogura et al. 2001, Helms et al. 2003, Edwards et al. 2005, Haines et al. 2005, Helms et al. 2005, Klein et al. 2005, Duerr et al. 2006, Grant et al. 2006, Nair et al. 2006, McPherson et al. 2007), we outline similar strategies for large-scale studies of IS to identify other common risk factors. IS may in fact provide interesting advantages in such studies, given that phenotyping involves standardized objective measures, and onset in childhood enables collection of entire families that can provide information on linkage, association, and inheritance. Identification of genetic associations in IS is likely to be rapid, but the functional relevance of these findings may emerge more slowly. We address this and the "elephant in the room" of scoliosis research, the need for a better understanding of normal adolescent spinal development. Model organisms are likely to be critically important to this challenge. Finally, we discuss the clinical implications of genetic findings in IS.

Population Studies

Familial and Twin Studies

Familial forms of IS were described as early as 1922 (Staub 1922) and have been well-documented in the literature since then (Garland 1934, Filho and Thompson

1971, Cowell et al. 1972, Riseborough and Wynne-Davies 1973, Robin and Cohen 1975, Wynne-Davies 1975, Bell and Teebi 1995). Reports of twin studies have also supported the genetic basis of IS (Horton 2002). Twin study designs compare the clinical similarity of monozygotic (MZ), or identical, twins who share 100% of their genes, to that of dizygotic (DZ,) or fraternal, twins who share only 50% of their genes. By studying and estimating concordance rates in many twin sets, the genetic effects of disease can be better understood. Most of the twin studies for IS consistently revealed higher concordances in monozygotic (MZ) compared to dizygotic twins. A meta-analysis of these clinical twin studies revealed 73% MZ concordances compared to 36% DZ concordances (Kesling and Reinker 1997). Interestingly, in this series when curve measurements were compared between monozygous twins, the correlation coefficient was significantly greater than the same correlation measurement between dizygous twins ($P < 0.0002$). No correlation with curve pattern was found, suggesting the importance of genetic factors in controlling susceptibility and disease course, but not necessarily disease pattern. A more recent report utilizing the Danish Twin Registry found 25% proband-wise concordance in monozygotic twins (6 of 44 concordant) compared to 0% concordance (0 of 91) in dizygotic twins, with an overall prevalence of approximately 1% (Andersen et al. 2007). The lower concordances in both groups as compared with prior results may be explained by differences in methodology, i.e., ascertainment in clinics vs. registry and screening by examination vs. questionnaire. Nevertheless, the overall trend obtained for all studies suggests strong genetic effects in IS. A second important feature of IS revealed by these studies was that monozygotic twins shared disease less than 100% of the time, reflecting the complexity of disease and suggesting the involvement of as yet unknown environmental or stochastic factors in disease susceptibility.

Sibling Risk Studies

Familial risk values may be used to estimate and compare the genetic effects across diseases. About one quarter of IS patients report a positive family history of disease (Wynne-Davies 1968). Prior sibling risk studies of IS have reported 19 and 11.5% of siblings affected for $\geq 10°$, $\geq 20°$ curves, respectively, compared to population frequencies of $\leq 2\%$ (Riseborough and Wynne-Davies 1973, Rogala et al. 1978, Lonstein 1994). The ratio of affected siblings: population frequency is called the "sibling risk ratio (λ_s)," a value that may be compared to roughly estimate the strength of genetic contributions between different diseases. A recent study (Wise et al. 2008) estimated IS $\lambda_s = 8$ when disease was defined as $\geq 10°$ curves, and increasing to $\lambda_s = 23$ for disease defined as $\geq 20°$ curves. These values indicate reasonably strong cumulative genetic effects that are comparable to those for other well-described complex genetic diseases such as rheumatoid arthritis (RA) ($\lambda_s = 2$–17), Crohn's disease (CD) ($\lambda_s = 10$), type 1 diabetes (T1D) ($\lambda_s = 15$), or psoriasis ($\lambda_s = 4$–11.5) (Wise et al. 2008).

Inheritance of IS Susceptibility

Is there an inheritance model that explains IS? An autosomal dominant pattern has been suggested (Staub 1922, Garland 1934, Filho and Thompson 1971, Cowell et al. 1972, Riseborough and Wynne-Davies 1973, Robin and Cohen 1975, Wynne-Davies 1975, Bell and Teebi 1995) from evaluation of single families or small family collections. X-linked dominant inheritance has been a prevailing theory to explain apparent lack of male–male transmission (Cowell et al. 1972), although this was disputed after re-evaluation of X-ray data from original study subjects (Riseborough and Wynne-Davies 1973, Horton 2002). Various studies have found that IS disease risk is much greater in first-degree relatives of a proband than for more distant relatives. In one example, a comprehensive population study (Riseborough and Wynne-Davies 1973) reported overall risk to first-degree relatives of 11% compared to 2.4% and 1.4% in second-degree and third-degree relatives. A "multifactorial" inheritance model has been suggested to explain this, where several to many genes may interact with as yet unknown environmental factors. Regarding the latter, it is interesting that some but not all studies have found advanced maternal age for mothers of probands with IS (De George and Fisher 1967, Filho and Thompson 1971, Riseborough and Wynne-Davies 1973, Wynne-Davies 1975). The general consensus gathered from all of this is that, while families with dominant inheritance may exist, IS is generally a "complex" genetic disease that is not easily explained by existing models and most often appears to be sporadic. Insights derived from studies of other so-called complex diseases could help us understand such situations. For example, *de novo* copy number variation was recently described in ~10% of sporadic cases of autism spectrum disorders, also described as complex genetic disease of childhood (Sebat et al. 2007). Future genomic analysis of IS, as discussed in detail below, may provide insight into this issue.

Ethnic Patterns

As noted, IS is observed in all major ethnic groups. Many studies have examined ethnic prevalences, and we have attempted to summarize relevant findings in Table 9.1. These results should be interpreted with caution, as variations may reflect different screening methodologies. However, we note with interest that, while most studies have found 1–3% of school age children affected if the usual definition of IS was applied (greater than 10° Cobb angle measurement), a much lower prevalence of IS (0.03%) was found in a Bantu-speaking population of African school children as compared to 2.5% of their Caucasian counterparts (Segil 1974). While most modern populations of the world are descendants of humans who migrated "out of Africa" much earlier (Jobling et al. 2004, Mellars 2006, Jacobs et al. 2008), Bantu-speaking populations are mainly restricted to Africa and only recently expanded (Quintana-Murci et al. 2008). Such observations might hint at common genetic variation predisposing to IS across world populations that arose secondary to selective processes early in Africa (Barreiro et al. 2008).

Table 9.1 Prevalence of scoliosis reported in different ethnic population groups

Study	Ethnic populations studied	Diagnostic criteria	Inferences	Measured prevalence	References
1	USA black vs. white population	$\geq 10°$	No difference between blacks and whites	Population over 14 years of age: 1.9%	Shands and Eisberg (1955)
2	USA black vs. white population adults aged 25–74 years	NA	Average prevalence in USA: 8.3% Higher in blacks, but not statistically significant	Blacks: 9.7% Whites: 8.1%	Carter and Haynes (1987)
3	African Bantu population vs. white population of Johannesburg	$\geq 10°$	Lower prevalence in Bantu population than whites	Bantu population: 0.03% White population: 2.5%	Segil (1974)
4	Lapps vs. non-Lapps from Norway	$\geq 10°$	Lower prevalence in Lapps	Lapps: 0.5% Non-Lapps: 1.3%	Skogland and Miller (1978)
5	Children from Canada in age group 12–14 years	$\geq 5°$	Prevalence decreased to 2.2 % for $\geq 10°$	4.5%	Rogala et al. (1978)
6	School children from Athens in Greece	Curve of at least 10°	Small curves equally distributed in both sexes but large curves were more common in girls	Girls: 4.6% Boys: 1.1%	Smyrnis et al. (1979)
7	Population from Malmo in southern Sweden	$\geq 10°$	Prevalence was 2.8 % for $\geq 5°$	Girls: 3.2% Boys: 0.5%	Willner and Uden (1982)
8	Children from England 10–14 years	$\geq 5°$	Prevalence was 0.8% for $\geq 10°$	3.9%	Dickson (1983)

9	Chinese vs. Malaysian and Indian girls from Singapore	$\geq 5°$	Higher rate in Chinese girls	Chinese girls
				  (11–12 years): 1.9%
				  (16–17 years): 3.5%
				Malay girls
				  (11–12 years): 1.5%
				  (16–17 years): 1.7%
				Indian girls
				  (11–12 years): 0.8%
				  (16–17 years): 1.7%
					Daruwalla et al. (1985)

Table 9.1 (continued)

Study	Ethnic populations studied	Diagnostic criteria	Inferences	Measured prevalence	References
10	Children from Canada ages 8–15 years	$\geq 5°$	Prevalence decreased to 1.8% for $\geq 10°$	6.2%	Morais et al. (1985)
11	Ancestry from Morocco vs. ancestry from Iraq and western Europe in Israel	Forward bending test and radiographic curve measurement grade A (mild scoliosis <20°) grade B (<20° but with pain) grade C (>20° with pain/no pain)	Higher prevalence in both sexes with parental origin from Iraq and western Europe	Grade A: 8.4–13% in males; 7.3–13% in females Grade B: 0.8–1.8% in males; 1.0–2.5% in females	Shohat et al. (1988)
12	School children from Japan	$\geq 15°$	Incidence of scoliosis increased linearly according to age	Fifth-grade students Boys: 0.07% Girls: 0.44% Junior high school students Boys: 0.25% Girls: 1.77%	Ohtsuka et al. (1988)
13	School children 6–14 years of age from Leeds region of England	$\geq 5°$	Prevalence was higher in girls and increased with age	2.7%	Stirling et al. (1996)
14	School children in northwestern and central Greece	$\geq 10°$	Majority (1.5%) were individuals with smaller curves (10–19°)	1.7%	Soucacos et al. (1997)
15	European vs. Polynesians (Maori and Pacific Islander) of New Zealand	NA	Polynesians had higher frequency of scoliosis but lower frequency of IS compared to Europeans	NA	Ratahi et al. (2002)

Molecular Genetics

Cytogenetic Studies

Most genetic diseases are explained by relatively subtle, submicroscopic DNA changes (Botstein and Risch 2003). However, rare cases with chromosomal alterations visible by karyotype analysis can effectively pinpoint causative genes. For example, rearrangements of chromosome 17 helped to more precisely localize the *NF1* gene responsible for neurofibromatosis (Ledbetter et al. 1989), and alterations of chromosome 15 pinpointed an imprinted region important in Prader–Willi and Angelman syndromes [reviewed in (Bittel and Butler 2005)]. Many chromosomal alterations with phenotypes that include scoliosis have been reported, although most do not appear to recapitulate idiopathic forms, i.e., without obvious vertebral anomalies or coexisting diagnoses [see for example, (Kulkarni et al. 2008)]. One family with IS and segregating a pericentric inversion of chromosome 8 has been reported (Szappanos et al. 1997). Using methods of chromosomal breakpoint mapping, Bashiardes et al. (Bashiardes et al. 2004) found that one end of this inversion disrupted the 8q11.2 gene encoding gamma-1-syntrophin (*SNTG1*), while the other end of the inversion occurred in a gene-free region of 8p23. Subsequent analysis of the *SNTG1* gene in 152 additional IS patients revealed an apparent mutation in DNA samples from 3 unrelated patients. These changes were not detected in screens of 480 healthy control individuals. These results suggested that rare mutations in *SNTG1* could occur in a small percentage of patients, but also left open the possibility that other nearby genes encoded on chromosome 8 could be important in IS.

Linkage Analyses

Statistical methods may be used to "link" polymorphic biologic markers with disease. Linkage studies may be conducted in a hypothesis-driven fashion centering on selected candidate molecules. Specifically, markers in candidate molecules are assayed, and the probability of linkage with disease is calculated. With regard to IS, early linkage studies were driven by the hypothesis that variation in molecular structures comprising the spine could be responsible for disease susceptibility. Polymorphisms in the collagen-encoding genes *COL1A1*, *COL1A2*, the fibrillin 1-encoding gene *FBN1*, and the elastin-encoding gene *ELN* were tested (i.e., "genotyped") in family collections, but the results did not reveal evidence of linkage (Carr et al. 1992, Miller et al. 1996). The subsequent availability of DNA sequence polymorphisms across the entire human genome has enabled more global searches for IS-linked genes, without the need to formulate prior hypotheses regarding disease etiology. The first such genome-wide linkage scan was performed in a single family segregating IS in three generations (Wise et al. 2000). Subsequent genome-wide scans of IS families and family collections have been reported (Chan et al. 2002, Salehi et al. 2002, Justice et al. 2003, Miller et al. 2005, Alden et al. 2006). These have produced evidence of linkage to several regions with designations in Mendelian Inheritance in Man (MIM), an official registry of genetic disease loci (Table 9.2).

Table 9.2 Reported linkages for IS. Regions proposed to harbor IS susceptibility genes are given by cytogenetic location, position relative to known polymorphisms, and MIM designation

Chromosomal region	MIM locus	Flanking loci	Candidate gene	References
6p		D6S1051–D6S1017		Wise et al. (2000); Miller et al. (2005)
6q		D6S1053–D6S1021		Miller et al. (2005)
8q	IS3	D8S1477–D8S279		Gao et al. (2007)
9q	IS4	D9S938–D9S934	CHD7	Miller et al. (2005); Ocaka et al. (2008)
10q		D10S1222–D10S212		Wise et al. (2000)
16q		D16S764–D16S2624		Miller et al. (2005)
17p	IS2	D17S974–D17S1294		Salehi et al. (2002)
17q	IS5	S17S1806–17qter		Ocaka et al. (2008)
18q		D18S1357–D18S1371		Wise et al. (2000)
19p	IS1	D19S1034		Chan et al. (2002); Alden et al. (2006)
Xq		GATA144D04–GATA172D05		Justice et al. (2003)

Additionally, genome-wide scans in a collection of 202 families produced varying results dependent on the stratification of the data. Without prior stratification by inheritance model or disease severity, suggestive results were reported for regions of chromosomes 1 and 6. However, with stratification other regions became more significant. The authors concluded that regions of chromosomes 6, 9, 16, and 17 are primary regions of interest warranting further analysis (Miller et al. 2005). A recent study (Ocaka et al. 2008) of families of British descent replicated and refined the chromosome 9-linked locus (IS4) and also identified a novel linked locus at chromosome 17q25.3-qtel (IS5). Thus linkage analysis suggests 9q31.2-34.2-encoded genes as promising candidates for further investigation in IS susceptibility.

Genetic Etiology of IS

Models Involving Neurologic Tissues

Idiopathic scoliosis is seemingly a musculoskeletal deformity, yet a neurological pathogenesis is the prevailing opinion to explain disease susceptibility (Herring 2002). This concept is derived from many independent observations. As noted, scoliotic deformity is often seen in association with neurologic/neuromuscular disease as noted. There is also a well-documented association between apparently isolated scoliosis and spinal cord anomaly such as a syrinx, a fluid-filled cavity in the spinal cord (or brainstem). Whether scoliosis is secondary to the anomaly of neurologic tissues, or whether both are secondary to some other lesion is unclear (Herring 2002). Another intriguing model suggests a central balancing dysfunction, implicating deficits in oculo-vestibular (visual/hearing) and proprioceptive function. This has been the subject of several investigations. As an example, one study measured otolith vestibulo-ocular response in IS. The otolith is a structure in the vestibular labyrinth of the inner ear that is sensitive to gravity and linear acceleration. The results showed significant left–right asymmetry in horizontal eye movements in IS patients compared to matched healthy control children; moreover the average of horizontal, but not vertical, eye movements was significantly different (greater) in IS patients than in controls. This supported a hypothesis of altered oculo-vestibular control mechanisms in IS, but did not necessarily distinguish cause from effect. However, the observation of scoliosis induced by otolith lesion in guinea pigs argues that the spinal deformity is a secondary effect [reviewed in (Lonstein 1994, Lowe et al. 2000, Wiener-Vacher and Mazda 1998, Rousie et al. 1999, Guo et al. 2006, Mallau et al. 2007)]. Horizontal gaze and scoliosis are also conspicuously associated in the rare "extreme" autosomal recessive disease *h*orizontal *g*aze *p*alsy with *p*rogressive *s*coliosis (HGPPS, MIM #607313). This disease is caused by homozygous loss-of-function mutations in the *ROBO3* gene encoding a transmembrane receptor controlling axon guidance. Brain imaging in HGPPS patients has confirmed that uncrossed motor and sensory axonal projections are present in the hindbrain. Remarkably, the only obvious clinical findings on physical examination of HGPPS patients are absent horizontal eye movements and severe progressive scoliosis (Jen et al. 2004). Again, it is tempting to speculate that scoliosis in these patients is somehow secondary to the profound axonal abnormalities of the hindbrain.

Other brain lesions have been associated with scoliosis. Short-wave electrolysis creating microscopic lesions in the pons or periaqueductal gray matter have produced experimental scoliosis in rats and rabbits with low efficiency (Lowe et al. 2000). Interestingly, removal of the pineal gland from chickens, fish, and bipedal rats produces scoliosis that resembles IS (Machida et al. 1993, Machida et al. 1997, Machida et al. 1999, Bagnall et al. 2001, Fjelldal et al. 2004). This effect is not seen in quadrupeds and suggests that bipedalism/upright posture provokes the deformity. How the brain and/or spinal cord could "talk" to the spinal column and produce scoliosis, is unknown. In the case of HGPPS patients, the authors hypothesized secondary abrogation of muscle tone and locomotion to explain the associated scoliosis (Jen et al. 2004). An effect on melatonin-mediated biochemical pathways has been invoked to explain the scoliosis produced in pinealectomized animals but remains unproven (Moreau et al. 2004). Limited genetic association studies of common variants in genes encoding the melatonin receptors (*MTNR1A* and *MTNR1B*) have not proven conclusive (Qiu et al. 2007, Qiu et al. 2008). Collectively, there is a persistent association between neurologic alteration and scoliosis that may apply to some cases of IS, but more comprehensive genetic, biochemical, and physiologic studies are required to clarify this relationship (Herring 2002).

Follow-up studies of chromosomal regions linked to IS (Table 9.2) have identified one candidate gene in the IS3 region of chromosome 8p encoding the chromodomain helicase DNA-binding protein 7 (CHD7) protein. Specifically, alleles of multiple single nucleotide polymorphisms (SNPs) spanning a portion of the *CHD7* gene were significantly linked and associated with disease in a cohort of 53 white American families of European descent. Presumably a variant in this region of the gene, or one of the SNP alleles themselves, provokes disease by altering CHD7 protein in some way. While the cellular function of CHD7 is not clear, its structural similarity to other chromodomain helicases suggests a possible role in gene transcription (Lalani et al. 2006). Loss-of-function coding mutations in *CHD7* are also known to cause the CHARGE syndrome of multiple developmental anomalies that can include scoliosis, suggesting etiologic overlap between CHARGE and IS. In this regard, it is also interesting that otolith deficiency has been noted in CHARGE patients (Delahaye et al. 2007). Milder variants leading to a relative reduction of CHD7 protein has been proposed to explain its association with IS susceptibility (Gao et al. 2007). In the mouse, embryonic tissues expressing *Chd7* mRNA include the developing organs involved in the CHARGE syndrome, primarily brain, structure of the ear, eyes, heart, and kidney (Bosman et al. 2005, Lalani et al. 2006). Expression in spinal cord was also observed at the E11 stage during murine development (Kim et al. 2008). Less is known about postnatal *CHD7* expression patterns, and indeed interpreting these findings is hampered by the current deficiency in our understanding of postnatal spinal developmental programmes.

Models Involving Other Spinal Tissues

Whether IS could result from primary lesions of other soft tissues comprising the spinal column, specifically spinal ligaments or paraspinous muscles, is unknown. Immunohistochemical analysis of the ligamentum flavum revealed some disarrangement of fibers in IS patients compared to controls (Miller et al. 1994), yet other histochemical and ultrastructural studies of cartilage, bone, and growth plate in IS patients have been inconclusive (Rogala et al. 1978). Some studies have shown decreased glycosaminoglycan content in the intervertebral discs of IS patients [reviewed in (Lowe et al. 2000)]. Paraspinous muscles on either "side" of the scoliotic curve in IS patients have shown comparative differences in hypertrophy and electromyographic signaling of type I fibers. Specifically, by these measures the muscles at the convexity of the curve display relative activation, a phenomenon that was explained as a compensatory response to curve progression (Lonstein 1994, Lowe et al. 2000). Many studies, mostly in Asian cohorts, have reported generalized osteopenia in IS patients (Cheng et al. 1999, Lee et al. 2005, Cheng et al. 2006). One study found that osteopenia of the femoral neck was a prognostic indicator of curve progression, with an odds ratio of 2.3 (Hung et al. 2005). Recently, a case–control association study (197 cases and 172 controls and replication in 222 cases and 288 controls) found that a polymorphism within the promoter region of the matrilin-1 gene (*MATN1*) is associated with IS susceptibility in a Chinese population (Chen et al. 2008). This gene encodes a member of von Willebrand factor A domain-containing protein family and is involved in the formation of filamentous networks in the extracellular matrices of various tissues (Deak et al. 1997). Further studies are required to determine whether genetic variation in the structural components of the spine itself influences susceptibility and/or disease progression in IS.

Discovering New Genetic Factors in Idiopathic Scoliosis

As described in this chapter, the genetic factors underlying IS are largely undiscovered at present. Emerging methods to efficiently re-sequence candidate regions or entire genomes to identify causative disease genetic variations are anticipated in the near future (Schuster 2008, Wold and Myers 2008). In the meantime, microchip-based methods are now enabling efficient genotyping on the order of one million polymorphisms in the genome of a single individual. Large-scale genotyping in IS populations is currently in progress and should afford considerably more power and efficiency to detect disease-associated variants than other genetic methods previously applied to IS gene discovery (Risch 2000, Carlson et al. 2004). Here we discuss application of new technologies to discovering IS risk alleles affecting susceptibility as well as disease course.

Discovering Genetic Risk Factors for IS Susceptibility

The strong genetic effects measured for IS portend success in association studies, given sufficiently large, well-characterized study cohorts as addressed below. Disease-associated DNA polymorphisms may be found by virtue of altered frequency in populations of IS affected cases as compared to a similar population of healthy control individuals. An alternative design is family-based and compares the frequencies of transmitted vs. untransmitted alleles from parents to affected offspring. The major benefit of this approach as compared to a case–control design is that parents are ideal population controls for their children, thereby removing the potential problem of bias due to population substructure (Chanock et al. 2007). Ascertaining two parents as controls is more costly but otherwise feasible for childhood-onset disease, making family-based designs attractive for genetic studies of IS. Both case–control designs and family-based designs should prove useful in genome-wide and follow-up targeted association studies of IS. The first issue to tackle in this regard is power and the ability to detect true associations. Power in genetic studies is in part a function of the risk posed by the disease allele, or other nearby proxy alleles. Although the genetic contribution of any undiscovered variant is at the moment unknown, the cumulative genetic effects measured for IS are encouraging, particularly as compared to the well-studied complex diseases described at the beginning of this chapter. Indeed, the IS cohort studied for *CHD7* association, albeit small, displayed genotype relative risks greater than two for several associated variants (Gao et al. 2007). Power is also a function of the assayed polymorphism in a given population, and the frequency of disease in that population (Gordon and Finch 2005). The relatively high overall frequency of IS (2–3%) in most populations supports the possibility that common genetic variants could increase disease risk. Nevertheless, genetic heterogeneity that will dilute positive signals is inevitable. One solution to overcoming this issue is to increase sample size. Power estimates suggest that studying at least 2,000 cases will provide sufficient power to detect disease alleles contributing a range of relative risks (Wise et al. 2008). Various stratification schemes also may be useful for increasing genetic homogeneity (and thereby increasing selected genetic signals). For example, datasets may be stratified by gender, age at onset, or curve severity in the proband. Regarding gender, male onset is typically later than female onset, coinciding with the male adolescent growth spurt, and disease progression continues much later in development, through Risser stage 5 of bone growth (Karol et al. 1993). Regarding severity, other linkage studies of IS have found differences when data were stratified by curve severity (Miller et al. 2005, Alden et al. 2006). Also, increasing estimates of λ_s were observed for curves greater than 20°, suggesting greater genetic effect for more progressive disease.

Genomic regions underlying rare disorders involving scoliosis may be important in "idiopathic" forms of the disease, as noted with the *CHD7* gene associated with both the CHARGE syndrome and IS (Gao et al. 2007). It is interesting that several of these "candidate" disorders involve known genomic imbalances (i.e., large deletions or duplications). For example, scoliosis is well-described as part of the phenotype in relatively rare duplication/deletion syndromes such as Charcot–Marie–Tooth disease (Sturtz et al. 1997), Prader–Willi syndrome (Holm and Laurnen 1981), Smith–Magenis syndrome (Greenberg et al. 1996), spinal muscular atrophy (Sucato 2007) and Di George syndrome (Bassett et al. 2005). It is interesting that commonly occurring, smaller copy number variations (CNVs) have been observed within the genomic regions involved in these diseases (Table 9.3). Thus it is important to consider CNVs, whether rare or common, as possible genetic variations that could contribute to risk of IS. Methods to detect CNV include array comparative genomic hybridization (CGH) and quantitative analysis of chip-based SNP genotyping (Pinkel and Albertson 2005, Colella et al. 2007). An important consideration here is the possibility that de novo CNVs may cause disease analogous to autism spectrum disorders as described earlier in this chapter. For any associated CNV it will be interesting to also consider the possibility that dosage (i.e., absolute copy number) may correlate with some aspect of disease, such as severity or onset, as has been observed, for example, with the CMT-associated 17p duplication (Sturtz et al. 1997).

Discovering Risk Factors for Disease Course

As discussed earlier in this chapter, genetic factors may be highly relevant to curve progression in IS. A central issue for clinicians is predicting whether patients with a certain initial curvature will progress and require intervention, or whether they will not progress significantly, so that no intervention is required. Several statistical methodologies enable use of phenotype and genotype data to investigate such questions. For example, survival analysis methods (Hosmer et al. 2008) with longitudinal data may be used to test whether some combination of input variables would predict outcome for different patients (e.g., those with a progressive form of disease as compared with those without). In these methods, input variables may include, for example, SNP genotypes, gender, ethnicity, and initial measures of age, curve magnitude, curve pattern, Risser sign. A logical outcome variable is curve magnitude at skeletal maturity. Also, recently developed mixture modeling techniques (Nagin 1999, Nagin and Tremblay 2001) ask whether a sample of patients' disease progression trajectories may be decomposed into homogeneous subgroups of trajectories (e.g., patients with rapid progression in curvature as compared with those with slow progression in curvature), and what input variables classify patients into the different subgroups. In this way it may be possible to identify combinations of genotypes and phenotypes that classify and predict IS outcomes.

Table 9.3 List of IS associated syndromes, chromosomal locations, and copy number variations

Disease phenotype	Chromosomal location of genomic imbalances	Type of variation	References	Studies showing CNV in or in vicinity of these regions
Charcot–Marie–Tooth disease type 4B2	11p15.4	Deletion	Senderek et al. (2003)	Iafrate et al. (2004)
Charcot–Marie–Tooth disease type 1A	17p12	Duplication	Lupski (1998)	de Vries et al. (2005); Sharp et al. (2005); McCarroll et al. (2006)
Prader–Willi/Angelman syndrome	15q11–13	Deletion/uniparental disomy	Ledbetter et al. (1982); Williams et al. (1989)	Iafrate et al. (2004); de Vries et al. (2005); Conrad et al. (2006); McCarroll et al. (2006)
Smith–Magenis syndrome	17p11.2	Deletion	Juyal et al. (1996); Lupski (1998)	Tuzun et al. (2005)
Spinal muscular atrophy	5q13.2	Deletion	Campbell et al. (1997)	Iafrate et al. (2004); Sebat et al. (2004); de Vries et al. (2005); Sharp et al. (2005)
DiGeorge syndrome	22q11.2	Deletion	Carlson et al. (1997); Edelmann et al. (1999)	Sharp et al. (2005); Conrad et al. (2006); McCarroll et al. (2006)
CHARGE Syndrome	8q12.1	Deletion	Vissers et al. (2004); OMIM	Iafrate et al. (2004)
Russell Silver Syndrome	7p11.2 and 11p15.5	Uniparental disomy; imprinting; epimutation	Reviewed in Abu-Amero et al. (2008)	Iafrate et al. (2004)

Understanding Discovered IS Genes in the Context of Adolescent Spinal Development

Spinal development during embryogenesis has been well-studied at cellular and molecular levels (Jessell 2000, Ulloa and Briscoe 2007, Dessaud et al. 2008), but this is lacking at later stages. Normal postnatal development of the human spine involves four general features: (1) proper osseous development to provide a mechanical axis for the appendicular skeleton and subsequent muscular development; (2) appropriate characteristic curves in the osseous spinal columns to support bipedal gait; (3) continuous symmetric growth (both vertical and axial); and (4) proper protection of neural elements (Labrom 2007). However, our understanding of the genetic mechanisms and molecular pathways associated with these features remains poor. Studies of mouse, zebrafish, and chicken may be useful in this regard, particularly the former two as they are more easily manipulated genetically. Regarding the latter, Taylor (1971) produced a highly inbred line of chickens that developed scoliosis in an apparently dominant fashion (Rucker et al. 1986). Unfortunately, this strain no longer exists (personal communication). More recently, zebrafish have been used to effectively study early skeletal development that may be extrapolated to humans. The short generation time, the large number of offspring and the early transparency of the embryos make these organisms ideal to study spinal development with genetic approaches [(Beattie et al. 2007) review]. Regarding the mouse, we note that heterozygous loss-of-function mutations in *CHD7* and its homolog *CHD2* both produce apparent spinal deformity in the mouse [K. Steel, E. Bosman, unpublished communication, and (Kulkarni et al. 2008)]. These results may encourage researchers to utilize common zebrafish or mouse laboratory strains for future genetic studies of discovered IS genes. Again, however, we expect that these efforts will be aided considerably as the developmental biology of postnatal periods, particularly the adolescent period in humans, becomes better understood.

Clinical Implications

In the near term, we expect that genomic studies will discover common factors affecting susceptibility to IS. These discoveries will evoke new questions that can formulate testable hypotheses regarding the etiology of IS. For example, how much do genetic factors increase risk of disease? Are genetic factors important modifiers of disease, i.e., do they affect outcome and response to treatment? Are there different genetic risks in different ethnic groups? How is the difference in male versus female risk explained? How do IS genes contribute to normal spinal development? What cell types are relevant in the disease process? Answering these questions will provide fascinating insights into important developmental periods of childhood and will enhance the prospect for alternative, less invasive therapies for spinal deformity.

Acknowledgments We thank Stuart Almond and Sarah Tune for their expert help with images.

References

Abu-Amero, S., Monk, D., Frost, J., Preece, M., Stanier, P., and Moore, G.E. (2008). The genetic aetiology of Silver–Russell syndrome. J. Med. Genet. 45: 193–199.
Alden, K.J., Marosy, B., Nzegwu, N., Justice, C.M., Wilson, A.F., and Miller, N.H. 2006. Idiopathic scoliosis: identification of candidate regions on chromosome 19p13. Spine 31:1815–1819.

Andersen, M.O., Thomsen, K., and Kyvik, K.O. 2007. Adolescent idiopathic scoliosis in twins: a population-based survey. Spine 32:927–930.

Bagnall, K.M., Beuerlein, M., Johnson, P., Wilson, J., Raso, V.J., and Moreau, M. 2001. Pineal transplantation after pinealectomy in young chickens has no effect on the development of scoliosis. Spine 26:1022–1027.

Barreiro, L.B., Laval, G., Quach, H., Patin, E., and Quintana-Murci, L. 2008. Natural selection has driven population differentiation in modern humans. Nat. Genet. 40:340–345.

Bashiardes, S., Veile, R., Allen, M., Wise, C.A., Dobbs, M., Morcuende, J.A., Szappanos, L., Herring, J.A., Bowcock, A.M., and Lovett, M. 2004. SNTG1, the gene encoding gamma1-syntrophin: a candidate gene for idiopathic scoliosis. Hum. Genet .115:81–89.

Bassett, A.S., Chow, E.W., Husted, J., Weksberg, R., Caluseriu, O., Webb, G.D., and Gatzoulis, M.A. 2005. Clinical features of 78 adults with 22q11 Deletion Syndrome. Am. J. Med. Genet. A 138:307–313.

Beattie, C.E., Carrel, T.L., and McWhorter, M.L. 2007. Fishing for a mechanism: using zebrafish to understand spinal muscular atrophy. J. Child Neurol. 22:995–1003.

Bell, M., and Teebi, A.S. 1995. Autosomal dominant idiopathic scoliosis? Am. J. Med. Genet. 55:112.

Bittel, D.C., and Butler, M.G. 2005. Prader-Willi syndrome: clinical genetics, cytogenetics and molecular biology. Expert Rev. Mol. Med. 7:1–20.

Bosman, E.A., Penn, A.C., Ambrose, J.C., Kettleborough, R., Stemple, D.L., and Steel, K.P. 2005. Multiple mutations in mouse Chd7 provide models for CHARGE syndrome. Hum. Mol. Genet. 14:3463–3476.

Botstein, D., and Risch, N. 2003. Discovering genotypes underlying human phenotypes: past successes for Mendelian disease, future approaches for complex disease. Nat. Genet. 33 Suppl:228–237.

Campbell, L., Potter, A., Ignatius, J., Dubowitz, V., and Davies, K. (1997). Genomic variation and gene conversion in spinal muscular atrophy: implications for disease process and clinical phenotype. Am. J. Hum. Genet. 61: 40–50.

Carlson, C.S., Eberle, M.A., Kruglyak, L., and Nickerson, D.A. 2004. Mapping complex disease loci in whole-genome association studies. Nature 429:446–452.

Carlson, C., Sirotkin, H., Pandita, R., Goldberg, R., McKie, J., Wadey, R., Patanjali, S.R., Weissman, S.M., Anyane-Yeboa, K., Warburton, D., Scrambler, P., Shprintzen, R., Kucherlapati, R., and Morrow, B.E. (1997). Molecular definition of 22q11 deletions in 151 vela-cardio-facial syndrome patients. Am. J. Hum. Genet. 61: 620–629.

Carr, A.J., Ogilvie, D.J., Wordsworth, B.P., Priestly, L.M., Smith, R., and Sykes, B. 1992. Segregation of structural collagen genes in adolescent idiopathic scoliosis. Clin. Orthop. Relat. Res. 274:305–310.

Carter, O.D., and Haynes, S.G. 1987. Prevalence rates for scoliosis in US adults: results from the first National Health and Nutrition Examination Survey. Int. J. Epidemiol. 16: 537–544.

Chan, V., Fong, G.C., Luk, K.D., Yip, B., Lee, M.K., Wong, M.S., Lu, D.D., and Chan, T.K. 2002. A genetic locus for adolescent idiopathic scoliosis linked to chromosome 19p13.3. Am. J. Hum. Genet. 71:401–406.

Chanock, S.J., Manolio, T., Boehnke, M., Boerwinkle, E., Hunter, D.J., Thomas, G., Hirschhorn, J.N., Abecasis, G., Altshuler, D., Bailey-Wilson, J.E., Brooks, L.D., Cardon, L.R., Daly, M., Donnelly, P., Fraumeni, J.F., Jr., Freimer, N.B., Gerhard, D.S., Gunter, C., Guttmacher, A.E., Guyer, M.S., Harris, E.L., Hoh, J., Hoover, R., Kong, C.A., Merikangas, K.R., Morton, C.C., Palmer, L.J., Phimister, E.G., Rice, J.P., Roberts, J., Rotimi, C., Tucker, M.A., Vogan, K.J., Wacholder, S., Wijsman, E.M., Winn, D.M., and Collins, F.S. 2007. Replicating genotype-phenotype associations. Nature 447:655–660.

Chen, Z., Tang, N.L., Cao, X., Qiao, D., Yi, L., Cheng, J.C., and Qiu, Y. 2008. Promoter polymorphism of matrilin-1 gene predisposes to adolescent idiopathic scoliosis in a Chinese population. Eur. J. Hum. Genet. 17(4):525–532.

Cheng, J.C., Guo, X., and Sher, A.H. 1999. Persistent osteopenia in adolescent idiopathic scoliosis. A longitudinal follow up study. Spine 24:1218–1222.

Cheng, J.C., Hung, V.W., Lee, W.T., Yeung, H.Y., Lam, T.P., Ng, B.K., Guo, X., and Qin, L. 2006. Persistent osteopenia in adolescent idiopathic scoliosis–longitudinal monitoring of bone mineral density until skeletal maturity. Stud. Health Technol. Inform. 123:47–51.

Colella, S., Yau, C., Taylor, J.M., Mirza, G., Butler, H., Clouston, P., Bassett, A.S., Seller, A., Holmes, C.C., and Ragoussis, J. 2007. QuantiSNP: an Objective Bayes Hidden-Markov Model to detect and accurately map copy number variation using SNP genotyping data. Nucleic Acids Res. 35:2013–2025.

Conrad, D.F., Andrews, T.D., Carter, N.P., Hurles, M.E., and Pritchard, J.K. 2006. A high-resolution survey of deletion polymorphism in the human genome. Nat. Genet, 38:75–81.

Cowell, H.R., Hall, J.N., and MacEwen, G.D. 1972. Genetic aspects of idiopathic scoliosis. A Nicholas Andry Award essay, 1970. Clin. Orthop. Relat. Res. 86:121–131.

Daruwalla, J.S., Balasubramaniam, P., Chay, S.O., Rajan, U., and Lee, H.P. 1985. Idiopathic scoliosis. Prevalence and ethnic distribution in Singapore schoolchildren. J. Bone Joint Surg. Br. 67:182–184.

De George, F.V., and Fisher, R.L. 1967. Idiopathic scoliosis: genetic and environmental aspects. J. Med. Genet. 4:251–257.

de Vries, B.B., Pfundt, R., Leisink, M., Koolen, D.A., Vissers, L.E., Janssen, I.M., Reijmersdal, S., Nillesen, W.M., Huys, E.H., Leeuw, N., Smeets, D., Sistermans, E.A., Feuth, T., van Ravenswaaij-Arts, C.M., van Kessel, A.G., Schoenmakers, E.F., Brunner, H.G., and Veltman, J.A. 2005. Diagnostic genome profiling in mental retardation. Am. J. Hum. Genet. 77:606–616.

Deak, F., Piecha, D., Bachrati, C., Paulsson, M., and Kiss, I. 1997. Primary structure and expression of matrilin-2, the closest relative of cartilage matrix protein within the von Willebrand factor type A-like module superfamily. J. Biol. Chem. 272:9268–9274.

Delahaye, A., Sznajer, Y., Lyonnet, S., Elmaleh-Berges, M., Delpierre, I., Audollent, S., Wiener-Vacher, S., Mansbach, A.L., Amiel, J., Baumann, C., Bremond-Gignac, D., Attie-Bitach, T., Verloes, A., and Sanlaville, D. 2007. Familial CHARGE syndrome because of CHD7 mutation: clinical intra- and interfamilial variability. Clin. Genet. 72:112–121.

Dessaud, E., McMahon, A.P., and Briscoe, J. 2008. Pattern formation in the vertebrate neural tube: a sonic hedgehog morphogen-regulated transcriptional network. Development. 135: 2489–2503.

Dickson, R.A. 1983. Scoliosis in the community. Br. Med. J. (Clin Res Ed). 286:615–618.

Doyle, C., and Blake, K. 2005. Scoliosis in CHARGE: a prospective survey and two case reports. Am. J. Med. Genet. A 133A:340–343.

Duerr, R.H., Taylor, K.D., Brant, S.R., Rioux, J.D., Silverberg, M..S, Daly, M.J., Steinhart, A.H., Abraham, C., Regueiro, M., Griffiths, A., Dassopoulos, T., Bitton, A., Yang, H., Targan, S., Datta, L.W., Kistner, E.O., Schumm, L.P., Lee, A.T., Gregersen, P.K., Barmada, M.M., Rotter, J.I., Nicolae, D.L., and Cho, J.H. 2006. A genome-wide association study identifies IL23R as an inflammatory bowel disease gene. Science 314:1461–1463.

Edelmann, L., Pandita, R.K., Spiteri, E., Funke, B., Goldberg, R., Palanisamy, N., Chaganti, R.S., Mgenis, E., Shprintzen, R.J., and Morrow, B.E. (1999). A common molecular basis for rearrangement disorders on chromosome 22q11. Hum. Mol. Genet. 8:1157–1167.

Edwards, A.O., Ritter, R., 3rd, Abel, K.J., Manning, A., Panhuysen, C., and Farrer, L.A. 2005. Complement factor H polymorphism and age-related macular degeneration. Science 308: 421–424.

Filho, N.A., and Thompson, M.W. 1971. Genetic studies in scoliosis. J. Bone Joint Surg. Am. 53:199.

Fjelldal, P.G., Grotmol, S., Kryvi, H., Gjerdet, N.R., Taranger, G.L., Hansen, T., Porter, M.J., and Totland, G.K. 2004. Pinealectomy induces malformation of the spine and reduces the mechanical strength of the vertebrae in Atlantic salmon, Salmo salar. J. Pineal Res. 36:132–139.

Gao, X., Gordon, D., Zhang, D., Browne, R., Helms, C., Gillum, J., Weber, S., Devroy, S., Swaney, S., Dobbs, M., Morcuende, J., Sheffield, V., Lovett, M., Bowcock, A., Herring, J., and Wise,

C. 2007. CHD7 gene polymorphisms are associated with susceptibility to idiopathic scoliosis. Am. J. Hum. Genet. 80:957–965.

Garland, H.G. 1934. Hereditary scoliosis. Br. Med. J. 1:328.

Gordon, D., and Finch, S.J. 2005. Factors affecting statistical power in the detection of genetic association. J. Clin. Invest. 115:1408–1418.

Grant, S.F., Thorleifsson, G., Reynisdottir, I., Benediktsson, R., Manolescu, A., Sainz, J., Helgason, A., Stefansson, H., Emilsson, V., Helgadottir, A., Styrkarsdottir, U., Magnusson, K.P., Walters, G.B., Palsdottir, E., Jonsdottir, T., Gudmundsdottir, T., Gylfason, A., Saemundsdottir, J., Wilensky, R.L., Reilly, M.P., Rader, D.J., Bagger, Y., Christiansen, C., Gudnason, V., Sigurdsson, G., Thorsteinsdottir, U., Gulcher, J.R., Kong, A., and Stefansson, K. 2006. Variant of transcription factor 7-like 2 (TCF7L2) gene confers risk of type 2 diabetes. Nat. Genet. 38:320–323.

Greenberg, F., Lewis, R.A., Potocki, L., Glaze, D., Parke, J., Killian, J., Murphy, M.A., Williamson, D., Brown, F., Dutton, R., McCluggage, C., Friedman, E., Sulek, M., and Lupski, J.R. 1996. Multi-disciplinary clinical study of Smith-Magenis syndrome (deletion 17p11.2). Am. J. Med. Genet. 62:247–254.

Guo, X., Chau, W.W., Hui-Chan, C.W., Cheung, C.S., Tsang, W.W., and Cheng, J.C. 2006. Balance control in adolescents with idiopathic scoliosis and disturbed somatosensory function. Spine 31:E437–440.

Haines, J.L., Hauser, M.A., Schmidt, S., Scott, W.K., Olson, L.M., Gallins, P., Spencer, K.L., Kwan, S.Y., Noureddine, M., Gilbert, J.R., Schnetz-Boutaud, N., Agarwal, A., Postel, E.A., and Pericak-Vance, M.A. 2005. Complement factor H variant increases the risk of age-related macular degeneration. Science 308:419–421.

Helms, C., Cao, L., Krueger, J.G., Wijsman, E.M., Chamian, F., Gordon, D., Heffernan, M., Daw, J.A., Robarge, J., Ott, J., Kwok, P.Y., Menter, A., and Bowcock, A.M. 2003. A putative RUNX1 binding site variant between SLC9A3R1 and NAT9 is associated with susceptibility to psoriasis. Nat. Genet. 35:349–356.

Helms, C., Saccone, N.L., Cao, L., Daw, J.A., Cao, K., Hsu, T.M., Taillon-Miller, P., Duan, S., Gordon, D., Pierce, B., Ott, J., Rice, J., Fernandez-Vina, M.A., Kwok, P.Y., Menter, A., and Bowcock, A.M. 2005. Localization of PSORS1 to a haplotype block harboring HLA-C and distinct from corneodesmosin and HCR. Hum. Genet. 118:466–476.

Herring, J.A. 2002. Tachdjian's Pediatric Orthopaedics, 3rd Ed. Philadelphia: W.B. Saunders Company

Holm, V.A., and Laurnen, E.L. 1981. Prader-Willi syndrome and scoliosis. Dev. Med. Child Neurol. 23:192–201.

Horton, D. 2002. Common skeletal deformities. In Emery & Rimoins Principles and Practices of Medical Genetics, eds. D.L. Rimoin , J.M. Connor, R.E. Pyeritz, and B.R. Korf, pp. 4236–4244. Amsterdam: Churchill Livingstone Elsevier

Hosmer, D.A., Lemeshow, S., and May, S. 2008. Applied Survival Analysis Regression Modeling of Time to Event Data. Wiley Series in Probability and Statistics. New York, NY: J. Wiley and Sons

Hung, V.W., Qin, L., Cheung, C.S., Lam, T.P., Ng, B.K., Tse, Y.K., Guo, X., Lee, K.M., and Cheng, J.C. 2005. Osteopenia: a new prognostic factor of curve progression in adolescent idiopathic scoliosis. J. Bone Joint Surg. Am. 87:2709–2716.

Iafrate, A.J., Feuk, L., Rivera, M.N., Listewnik, M.L., Donahoe, P.K., Qi, Y., Scherer, S.W., and Lee, C. 2004. Detection of large-scale variation in the human genome. Nat. Genet. 36:949–951.

Jacobs, Z., Roberts, R.G., Galbraith, R.F., Deacon, H.J., Grun, R., Mackay, A., Mitchell, P., Vogelsang, R., and Wadley, L. 2008. Ages for the Middle Stone Age of southern Africa: implications for human behavior and dispersal. Science 322:733–735.

James, J.I. 1954. Idiopathic scoliosis; the prognosis, diagnosis, and operative indications related to curve patterns and the age at onset. J. Bone Joint Surg. Br. 36-B:36–49.

Jen, J.C., Chan, W.M., Bosley, T.M., Wan, J., Carr, J.R., Rub, U., Shattuck, D., Salamon, G., Kudo, L.C., Ou, J., Lin, D.D., Salih, M.A., Kansu, T., Al Dhalaan, H., Al Zayed, Z., MacDonald, D.B.,

Stigsby, B., Plaitakis, A., Dretakis, E.K., Gottlob, I., Pieh, C., Traboulsi, E.I., Wang, Q., Wang, L., Andrews, C., Yamada, K., Demer, J.L., Karim, S., Alger, J.R., Geschwind, D.H., Deller, T., Sicotte, N.L., Nelson, S.F., Baloh, R.W., and Engle, E.C. 2004. Mutations in a human ROBO gene disrupt hindbrain axon pathway crossing and morphogenesis. Science. 304:1509–1513.

Jessell, T.M. 2000. Neuronal specification in the spinal cord: inductive signals and transcriptional codes. Nat. Rev. Genet. 1:20–29.

Jobling, M.A., Hurles, M.E., and Tyler-Smith, C. 2004. Human Evolutionary Genetics: Origins, Peoples, and Disease. New York, NY: Garland

Justice, C.M., Miller, N.H., Marosy, B., Zhang, J., and Wilson, A.F. 2003. Familial idiopathic scoliosis: evidence of an X-linked susceptibility locus. Spine 28:589–594.

Juyal, R.C., Figuera, L.E., Hauge, X., Elsea, S.H., Lupski, J.R., Greenberg, F., Baldini, A., and Patel, P.I. (1996). Molecular analyses of 17p11.2 deletions in 62 Smith-Magenis syndrome patients. Am. J. Hum. Genet. 58:998–1007.

Karol, L.A., Johnston, C.E., 2nd, Browne, R.H., and Madison, M. 1993. Progression of the curve in boys who have idiopathic scoliosis. J. Bone Joint Surg. Am. 75:1804–1810.

Kesling, K.L., and Reinker, K.A. 1997. Scoliosis in twins. A meta-analysis of the literature and report of six cases. Spine. 22:2009–2014.

Kim, H.G., Kurth, I., Lan, F., Meliciani, I., Wenzel, W., Eom, S.H., Kang, G.B., Rosenberger, G., Tekin, M., Ozata, M., Bick, D.P., Sherins, R.J., Walker, S.L., Shi, Y., Gusella, J.F., and Layman, L.C. 2008. Mutations in CHD7, encoding a chromatin-remodeling protein, cause idiopathic hypogonadotropic hypogonadism and Kallmann syndrome. Am. J. Hum. Genet. 83: 511–519.

Klein, R.J., Zeiss, C., Chew, E.Y., Tsai, J.Y., Sackler, R.S., Haynes, C., Henning, A.K., SanGiovanni, J.P., Mane, S.M., Mayne, S.T., Bracken, M.B., Ferris, F.L., Ott, J., Barnstable, C., and Hoh, J. 2005. Complement factor H polymorphism in age-related macular degeneration. Science 308:385–389.

Kulkarni, S., Nagarajan, P., Wall, J., Donovan, D.J., Donell, R.L., Ligon, A.H., Venkatachalam, S., and Quade, B.J. 2008. Disruption of chromodomain helicase DNA binding protein 2 (CHD2) causes scoliosis. Am, J. Med. Genet, A. 146A:1117–1127.

Labrom, R.D. 2007. Growth and maturation of the spine from birth to adolescence. J. Bone Joint Surg. Am. 89 Suppl 1:3–7.

Lalani, S.R., Safiullah, A.M., Fernbach, S.D., Harutyunyan, K.G., Thaller, C., Peterson, L.E., McPherson, J.D., Gibbs, R.A., White, L.D., Hefner, M., Davenport, S.L., Graham, J.M., Bacino, C.A., Glass, N.L., Towbin, J.A., Craigen, W.J., Neish, S.R., Lin, A.E., and Belmont, J.W. 2006. Spectrum of CHD7 mutations in 110 individuals with CHARGE syndrome and genotype-phenotype correlation. Am. J. Hum. Genet. 78:303–314.

Lupski, J.R. (1998). Genomic disorders: structural features of the genome can lead to DNA rearrangements and human disease traits. Trends Genet. 14:417–422.

Ledbetter, D.H., Rich, D.C., O'Connell, P., Leppert, M., and Carey, J.C. 1989. Precise localization of NF1 to 17q11.2 by balanced translocation. Am. J. Hum. Genet. 44:20–24.

Lee, W.T., Cheung, C.S., Tse, Y.K., Guo, X., Qin, L., Lam, T.P., Ng, B.K., Cheng, J.C. 2005. Association of osteopenia with curve severity in adolescent idiopathic scoliosis: a study of 919 girls. Osteoporos. Int. 16:1924–1932.

Lonstein, J.E. 1994. Adolescent idiopathic scoliosis. Lancet. 344:1407–1412.

Lowe, T.G., Edgar, M., Margulies, J.Y., Miller, N.H., Raso, V.J., Reinker, K.A., and Rivard, C.H. 2000. Etiology of idiopathic scoliosis: current trends in research. J. Bone Joint Surg. Am. 82-A:1157–1168.

Lupski, J.R. (1998). Genomic disorders: structural features of the genome can lead to DNA rearrangements and human disease traits. Trends Genet. 14:417–422.

Machida, M., Dubousset, J., Imamura, Y., Iwaya, T., Yamada, T., and Kimura, J. 1993. An experimental study in chickens for the pathogenesis of idiopathic scoliosis. Spine 18:1609–1615.

Machida, M., Miyashita, Y., Murai, I., Dubousset, J., Yamada, T., and Kimura, J. 1997. Role of serotonin for scoliotic deformity in pinealectomized chicken. Spine 22:1297–1301.

Machida, M., Murai, I., Miyashita, Y., Dubousset, J., Yamada, T., and Kimura, J. 1999. Pathogenesis of idiopathic scoliosis. Experimental study in rats. Spine 24:1985–1989.

Mallau, S., Bollini, G., Jouve, J.L., and Assaiante, C. 2007. Locomotor skills and balance strategies in adolescents idiopathic scoliosis. Spine 32:E14–E22.

McCarroll, S.A., Hadnott, T.N., Perry, G.H., Sabeti, P.C., Zody, M.C., Barrett, J.C., Dallaire, S., Gabriel, S.B., Lee, C., Daly, M.J., and Altshuler, D.M. 2006. Common deletion polymorphisms in the human genome. Nat. Genet. 38:86–92.

McPherson, R., Pertsemlidis, A., Kavaslar, N., Stewart, A., Roberts, R., Cox, D.R., Hinds, D.A., Pennacchio, L.A., Tybjaerg-Hansen, A., Folsom, A.R., Boerwinkle, E., Hobbs, H.H., and Cohen, J.C. 2007. A common allele on chromosome 9 associated with coronary heart disease. Science 316:1488–1491.

Mellars, P. 2006. Why did modern human populations disperse from Africa ca. 60,000 years ago? A new model. Proc. Natl. Acad. Sci. U. S. A. 103:9381–9386.

Miller, N.H., Justice, C.M., Marosy, B., Doheny, K.F., Pugh, E., Zhang, J., Dietz, H.C., 3rd, and Wilson, A.F. 2005. Identification of candidate regions for familial idiopathic scoliosis. Spine 30:1181–1187.

Miller, N.H., Mims, B., Child, A., Milewicz, D.M., Sponseller, P., and Blanton, S.H. 1996. Genetic analysis of structural elastic fiber and collagen genes in familial adolescent idiopathic scoliosis. J. Orthop. Res. 14:994–999.

Miller, N.H., Mims, B., and Milewicz, D.M. 1994. The potential role of elastic fiber system in adolescent idiopathic scoliosis. J. Bone Joint Surg. Am. 76:1193–1206.

Morais, T., Bernier, M., and Turcotte, F. 1985. Age- and sex-specific prevalence of scoliosis and the value of school screening programs. Am. J. Public Health. 75:1377–1380.

Moreau, A., Wang, D.S., Forget, S., Azeddine, B., Angeloni, D., Fraschini, F., Labelle, H., Poitras, B., Rivard, C.H., and Grimard, G. 2004. Melatonin signaling dysfunction in adolescent idiopathic scoliosis. Spine 29:1772–1781.

Nagin, D. 1999. Analyzing developmental trajectories: a semi-parametric, group-based approach. Psychol. Methods 4:139–177.

Nagin, D.S., and Tremblay, R.E. 2001. Analyzing developmental trajectories of distinct but related behaviors: a group-based method. Psychol. Methods 6:18–34.

Nair, R.P., Stuart, P.E., Nistor, I., Hiremagalore, R., Chia, N.V., Jenisch, S., Weichenthal, M., Abecasis, G.R., Lim, H.W., Christophers, E., Voorhees, J.J., and Elder, J.T. 2006. Sequence and haplotype analysis supports HLA-C as the psoriasis susceptibility 1 gene. Am. J. Hum. Genet. 78:827–851.

Ocaka, L., Zhao, C., Reed, J.A., Ebenezer, N.D., Brice, G.,,. Morley, T., Mehta, M., O'Dowd, J., Weber, J.L., Hardcastle, A.J., and Child, A.H. 2008. Assignment of two loci for autosomal dominant adolescent idiopathic scoliosis to chromosomes 9q31.2-q34.2 and 17q25.3-qtel. J. Med. Genet. 45:87–92.

Ogura, Y., Bonen, D.K., Inohara, N., Nicolae, D.L., Chen, F.F., Ramos, R., Britton, H., Moran, T., Karaliuskas, R., Duerr, R.H., Achkar, J.P., Brant, S.R., Bayless, T.M., Kirschner, B.S., Hanauer, S.B., Nunez, G., and Cho, J.H. 2001. A frameshift mutation in NOD2 associated with susceptibility to Crohn's disease. Nature 411:603–606.

Ohtsuka, Y., Yamagata, M., Arai, S., Kitahara, H.,and Minami, S. 1988. School screening for scoliosis by the Chiba University Medical School screening program. Results of 1.24 million students over an 8-year period. Spine 13:1251–1257.

Pehrsson, K., Bake, B., Larsson, S., and Nachemson, A. 1991. Lung function in adult idiopathic scoliosis: a 20 year follow up. Thorax 46:474–478.

Pinkel, D., and Albertson, D.G. 2005. Array comparative genomic hybridization and its applications in cancer. Nat. Genet. 37(Suppl):S11–S17.

Qiu, X.S., Tang, N.L., Yeung, H.Y., Cheng, J.C., and Qiu, Y. 2008. Lack of association between the promoter polymorphism of the MTNR1A gene and adolescent idiopathic scoliosis. Spine 33:2204–2207.

Qiu, X.S., Tang, N.L., Yeung, H.Y., Lee, K.M., Hung, V.W., Ng, B.K., Ma, S.L., Kwok, R.H., Qin, L., Qiu, Y., and Cheng, J.C. 2007. Melatonin receptor 1B (MTNR1B) gene polymorphism is associated with the occurrence of adolescent idiopathic scoliosis. Spine 32:1748–1753.

Quintana-Murci, L., Quach, H., Harmant, C., Luca, F., Massonnet, B., Patin, E., Sica, L., Mouguiama-Daouda, P., Comas, D., Tzur, S., Balanovsky, O., Kidd, K.K., Kidd, J.R., van der Veen, L., Hombert, J.M., Gessain, A., Verdu, P., Froment, A., Bahuchet, S., Heyer, E., Dausset, J., Salas, A., and Behar, D.M. 2008. Maternal traces of deep common ancestry and asymmetric gene flow between Pygmy hunter-gatherers and Bantu-speaking farmers. Proc. Natl. Acad. Sci. U. S. A. 105:1596–1601.

Ratahi, E.D., Crawford. H.A., Thompson. J.M., and Barnes, M.J. 2002. Ethnic variance in the epidemiology of scoliosis in New Zealand. J. Pediatr. Orthop. 22:784–787.

Richards, B.S., and Vitale, M.G. 2008. Screening for idiopathic scoliosis in adolescents. An information statement. J. Bone Joint Surg. Am. 90:195–198.

Risch, N.J. 2000. Searching for genetic determinants in the new millennium. Nature 405:847–856.

Riseborough, E.J., and Wynne-Davies, R. 1973. A genetic survey of idiopathic scoliosis in Boston, Massachusetts. J. Bone Joint Surg. Am. 55:974–982.

Robin, G.C., and Cohen, T. 1975. Familial scoliosis. A clinical report. J. Bone Joint Surg. Br. 57:146–148.

Rogala, E.J., Drummond, D.S., and Gurr, J. 1978. Scoliosis: incidence and natural history. A prospective epidemiological study. J. Bone Joint Surg. Am. 60:173–176.

Rousie, D., Hache, J.C., Pellerin, P., Deroubaix, J.P., Van Tichelen, P., and Berthoz, A. 1999. Oculomotor, postural, and perceptual asymmetries associated with a common cause. Craniofacial asymmetries and asymmetries in vestibular organ anatomy. Ann .N. Y. Acad. Sci. 871:439–446.

Rucker, R., Opsahl, W., Abbott, U., Greve, C., Kenney, C., and Stern, R. 1986. Scoliosis in chickens. A model for the inherited form of adolescent scoliosis. Am. J. Pathol. 123:585–588.

Salehi, L.B., Mangino, M., De Serio, S., De Cicco, D., Capon, F., Semprini, S., Pizzuti, A., Novelli, G., and Dallapiccola, B. 2002. Assignment of a locus for autosomal dominant idiopathic scoliosis (IS) to human chromosome 17p11. Hum. Genet. 111:401–404.

Schuster, S.C. 2008. Next-generation sequencing transforms today's biology. Nat. Methods 5: 16–18.

Sebat, J., Lakshmi, B., Malhotra, D., Troge, J., Lese-Martin, C., Walsh, T., Yamrom, B., Yoon, S., Krasnitz, A., Kendall, J., Leotta, A., Pai, D., Zhang, R., Lee, Y.H., Hicks, J., Spence, S.J., Lee, A.T., Puura, K., Lehtimaki, T., Ledbetter, D., Gregersen, P.K., Bregman, J., Sutcliffe, J.S., Jobanputra, V., Chung, W., Warburton, D., King, M.C., Skuse, D. Geschwind, D.H., Gilliam, T.C., Ye, K., and Wigler, M. 2007. Strong association of de novo copy number mutations with autism. Science 316:445–449.

Sebat, J., Lakshmi, B., Troge, J., Alexander, J., Young, J., Lundin, P., Maner, S., Massa, H., Walker, M., Chi, M., Navin, N., Lucito, R., Healy, J., Hicks, J., Ye, K., Reiner, A., Gilliam, T.C., Trask, B., Patterson, N., Zetterberg, A., and Wigler, M. 2004. Large-scale copy number polymorphism in the human genome. Science 305:525–528

Segil, C.M. 1974. The incidence of idiopathic scoliosis in the Bantu and White population groups in Johannesburg. J. Bone Joint Surg. Br. 56:393.

Senderek, J., Bergmann, C., Weber, S., Ketelsen, U.P., Schorle, H., Rudnik-schoneborn, S., Buttner, R., Buchheim, E., and Zerres, K. (2003). Mutation of the SBF2 gene, encoding a novel member of the myotubularin family, in Charcot-Marie-Tooth neuropathy type 4B2/ 11p15. Hum. Mol. Genet. 12:349–356

Shands, A.R., Jr., and Eisberg, H.B. 1955. The incidence of scoliosis in the state of Delaware; a study of 50,000 minifilms of the chest made during a survey for tuberculosis. J. Bone Joint Surg. Am. 37-A:1243–1249.

Sharp, A.J., Locke, D.P., McGrath, S.D., Cheng, Z., Bailey, J.A., Vallente, R.U., Pertz, L.M., Clark, R.A., Schwartz, S., Segraves, R., Oseroff, V.V., Albertson, D.G., Pinkel, D., and Eichler, E.E.

2005. Segmental duplications and copy-number variation in the human genome. Am. J. Hum. Genet. 77:78–88.

Shohat, M., Shohat, T., Nitzan, M., Mimouni, M., Kedem, R., and Danon, Y.L. 1988. Growth and ethnicity in scoliosis. Acta Orthop. Scand. 59:310–313.

Skogland, L.B., and Miller, J.A.A. 1978. The incidence of scoliosis in northern Norway. Acta Orthop. Scand. 49:635.

Smyrnis, P.N., Valavanis, J., Alexopoulos, A., Siderakis, G., and Giannestras, N.J. 1979. School screening for scoliosis in Athens. J. Bone Joint Surg. Br. 61-B:215–217.

Soucacos, P.N., Soucacos, P.K., Zacharis, K.C., Beris, A.E., and Xenakis, T.A. 1997. School-screening for scoliosis. A prospective epidemiological study in northwestern and central Greece. J. Bone Joint Surg. Am. 79:1498–1503.

Staub, H.A. 1922. Eine skoliotikerfamilie. Ein Beitrag zur Frage der kongenitalen Skoliose und der Hereditat der Skoliosen. Z. Orthop. Chir. 43:1.

Stirling, A.J., Howel, D., Millner, P.A., Sadiq, S., Sharples, D., and Dickson, R.A. 1996. Late-onset idiopathic scoliosis in children six to fourteen years old. A cross-sectional prevalence study. J. Bone Joint Surg. Am. 78:1330–1336.

Sturtz, F.G., Latour, P., Mocquard, Y., Cruz, S., Fenoll, B., LeFur, J.M., Mabin, D., Chazot, G., and Vandenberghe, A. 1997. Clinical and electrophysiological phenotype of a homozygously duplicated Charcot-Marie-Tooth (type 1A) disease. Eur. Neurol. 38:26–30.

Sucato, D.J. 2007. Spine deformity in spinal muscular atrophy. J. Bone Joint Surg. Am. 89 Suppl 1:148–154.

Szappanos, L., Balogh, E., Szeszak, F., Olah, E., Nagy, Z., and Szepesi, K. 1997. Idiopathic scoliosis–new surgical methods or search for the reasons. Acta Chir. Hung. 36:343–345.

Taylor, L.W. 1971. Kyphoscoliosis in a long-term selection experiment with chickens. Avian Dis. 15:376–390.

Tuzun, E., Sharp, A.J., Bailey, J.A., Kaul, R., Morrison, V.A., Pertz, L.M., Haugen, E., Hayden, H., Albertson, D., Pinkel, D., Olson, M.V., and Eichler, E.E. 2005. Fine-scale structural variation of the human genome. Nat. Genet. 37:727–732.

Ulloa, F., and Briscoe, J. 2007. Morphogens and the control of cell proliferation and patterning in the spinal cord. Cell Cycle 6:2640–2649.

Vissers, L.E., van Ravenswaaij, C.M., Admiraal, R., Hurst, J.A., de vries, B.B., Janssen, I.M., van der vliet, W.A., Huys, E.H., de Jong, P.J., Hamel, B.C., Schoenmakers, E.F., Brunner, H.G., Veltman, J.A., and van Kessel, A.G. (2004). Mutations in a new member of the chromodomain gene family cause CHARGE syndrome. Nat. Genet. 36:955–957

Wiener-Vacher, S.R., and Mazda, K. 1998. Asymmetric otolith vestibulo-ocular responses in children with idiopathic scoliosis. J. Pediatr. 132:1028–1032.

Williams, C.A., Gray, B.A., Hendrickson, J.E., Stone, J.W., and Cantu, E.S. (1989). Incidence of 15q deletions in the Angelman syndrome: a survey of twelve affected persons. Am. J. Med. Genet. 32:339–345.

Willner, S., and Uden, A. 1982. A prospective prevalence study of scoliosis in southern Sweden. Acta Orthop. Scand. 53:233–237.

Wise, C.A., Barnes, R., Gillum, J., Herring, J.A., Bowcock, A.M., and Lovett, M. 2000. Localization of susceptibility to familial idiopathic scoliosis. Spine 25:2372–2380.

Wise, C.A., Gao, X., Shoemaker, S., Gordon, D., and Herring, J.A. 2008. Understanding genetic factors in idiopathic scoliosis, a complex disease of childhood. Current Genomics 9:51–59.

Wold, B., and Myers, R.M. 2008. Sequence census methods for functional genomics. Nat. Methods. 5:19–21.

Wynne-Davies, R. 1968. Familial (idiopathic) scoliosis. A family survey. J. Bone Joint Surg. Br. 50:24–30.

Wynne-Davies, R. 1975. Infantile idiopathic scoliosis. Causative factors, particularly in the first six months of life. J. Bone Joint Surg. Br. 57:138–141.

Yamada, K., Miyamoto, K., Hosoe, H., Mizutani, M., and Shimizu, K. 2007. Scoliosis associated with Prader-Willi syndrome. Spine J. 7:345–348.

Chapter 10
Conclusion: Trends and Predictions for Genetic and Developmental Biological Research on Scoliosis

Kenro Kusumi

Introduction

It has been 5 years since the gold-standard, "finished" sequence has been released as part of the Human Genome Project (International Human Genome Sequencing Consortium 2004). The bioinformatic analysis yielded the surprising results that there may be only 20,000–25,000 protein-coding genes in the human genome, which is comparable to the number of genes in many "less complex" model systems. Concurrent genome sequencing efforts have been carried out for an ever-increasing number of model organisms used in scoliosis studies, including but not limited to the mouse, chick, zebrafish, and frog *Xenopus tropicalis*. Developmental genetic studies in these organisms have greatly expanded our understanding of the mechanisms regulating spinal development. The disruptions of these processes give us clues to the etiology of human spinal curves. Despite the advances in our understanding of somitogenesis and the earliest steps in spinal development, which are detailed further in Chapter 1, many questions remain about later developmental steps, as is clear from Chapter 2.

In this volume, the following topics on the current state of the genetics and development of scoliosis have been described:

1. *Developmental genetics of somitogenesis and early spinal patterning (Chapters 1 and 3):* Our current understanding of the genetic regulation of somitogenesis and the earliest steps in spinal development has derived from studies of animal models. These model systems will also allow us to better understand the environmental factors that may contribute to congenital scoliosis. The identification of genes involved in spinal development will help in the validation of gene candidates for ongoing human genetic efforts.

K. Kusumi (✉)
School of Life Sciences, Arizona State University, Tempe, AZ 85287, USA; Department of Basic Medical Sciences, The University of Arizona College of Medicine–Phoenix in Partnership with Arizona State University, Phoenix, AZ 85004, USA
e-mail: kenro.kusumi@asu.edu

K. Kusumi, S.L. Dunwoodie (eds.), *The Genetics and Development of Scoliosis*, 191
DOI 10.1007/978-1-4419-1406-4_10, © Springer Science+Business Media, LLC 2010

2. *Functional anatomy and later development of the spine (Chapter 2)*: While somitogenesis lays down the initial pattern of the spine, subsequent developmental events shape the formation of the vertebrae, axial muscle groups, tendons, and ligaments that are relevant for scoliosis. Genetic studies have identified regulatory pathways involved in the differentiation of paraxial mesodermal cells into embryonic muscle. Functional anatomical studies of normal and scoliotic patients have identified the key muscle groups that are required for spinal stability and flexion. However, the genetic regulation of later spinal development remains poorly understood.

3. *Classification of scoliosis for genetic studies (Chapters 4–7)*: Scoliosis and related spinal curves are features of numerous syndromic and non-syndromic disorders. Major categories include congenital and neuromuscular scoliosis as well as idiopathic scoliosis, which is the most commonly observed type. Collaborative groups of clinical geneticists, orthopedic surgeons, and genetics researchers are working to devise new classifications for scoliosis and related spinal curve disorders that reflect molecular and cellular mechanisms of disease.

4. *Human genetic analysis of congenital and idiopathic scoliosis*: Identification of segmental defects of the vertebrae (SDV) in pedigrees and linkage analysis studies have identified four genes (*DLL3, MESP2, LFNG,* and *HES7*) that are associated with congenital spinal deformities (Chapter 5). These genes have also been shown to play key roles in the genetic regulation of somitogenesis in animal models (Chapter 1). However, most cases of congenital and idiopathic scoliosis appear to display complex patterns of inheritance, and progress will depend on the assembly of study populations and execution of genome-wide association (GWA) studies and high-throughput resequencing efforts (Chapters 8 and 9). These efforts have identified the first locus associated with idiopathic scoliosis (*CHD7*), with the promise of many more genes identified in the upcoming years.

5. *Translational medicine (Chapters 4–9):* The ultimate goal of this research is to apply genetic and developmental knowledge to improve therapeutic intervention for all forms of scoliosis, and for idiopathic scoliosis, the early genetic diagnostic identification of the subpopulation prone to progression toward severe spinal curves.

The chapters in this volume make clear that we are at the cusp of a period of rapid progress in scoliosis research. What are the factors that are shaping future directions in scoliosis research? In this chapter, we will highlight the following technologies and research initiatives that will significantly impact on this field:

1. Technological advances and decreased cost of high-throughput sequencing
2. Establishment of research consortia focused on scoliosis
3. Building large-scale patient databases useful for scoliosis research
4. Future use of animal and cell culture models for scoliosis research
5. Utility of systems biology and bioinformatic approaches for scoliosis research

1. Technological Advances and Decreased Cost of High-Throughput Sequencing: The $1,000 Personal Genome

While Sanger technology-based approaches have been used to carry out whole genome shotgun sequencing efforts for the human and most animal models, next generation technologies hold great promise of reducing the cost and the time necessary for genome sequencing. Currently available technologies include the Roche 454, which uses a bead-based emulsion PCR for amplification and pyrosequencing (light producing reaction) technology, and the Illumina/Solexa system. The throughput is estimated to be approximately 1–1.5 gigabase-pair (Gbp) per day (reviewed in Turner et al. 2009). Other technologies include the Applied Biosystems SOLiD and Helicos tSMS systems, which have a slightly higher throughput of approximately 2 Gbp/day. These technologies generate relatively short sequence reads, which increase the bioinformatic challenge of sequence assembly (see below).

Beyond the current available systems, the next wave of sequencing technologies includes the use of nanopore technologies where the DNA is translocated through a membrane pore with detection of each nucleotide (Turner et al. 2009). The goal of these technological efforts is to produce the $1,000 genome, which would bring whole genome sequencing on par with costs of genotyping using microarray technologies. The availability of whole genome sequence for scoliosis cases raises both possibilities and challenges. With all available sequence polymorphisms, association studies could be carried out with a substantially larger group of genes than even the 100K SNP chips. However, the bioinformatic challenge of processing whole genome sequences of thousands of patients in a database requires substantial infrastructure, both storage capacity and trained personnel. It is possible that given the infrastructural requirements for such efforts, there will be consolidation of centers carrying out whole genome sequencing and analysis as a "service," akin to current sequencing or primer synthesis technologies. As will be discussed below, it is clear that as the cost decreases and the capacity of sequencing/genotyping technologies increases, the challenge is to screen larger patient databases.

2. Establishment of Research Consortia Focused on Scoliosis: Economies of Scale for the Next Decade

Since scoliosis is a diverse group of disorders, genetic studies have been hampered by the lack of a large patient database. However, efforts are underway to establish patient databases as described below.

Combining the resources of individual investigators or institutions is one clear approach to developing economies of scale for genetic research. One example of this approach is the International Consortium for Vertebral Anomalies and Scoliosis (ICVAS, http://www.icvas.org), which was founded in 2006 by clinicians and researchers committed to identifying the genetic and developmental mechanisms underlying scoliosis. ICVAS, as with many consortia, seeks to build a DNA database together with relevant clinical data from a diverse group of patients with spinal curvatures and vertebral anomalies. This database would allow investigators to carry out genome-wide association studies on a larger group of patients than would be available to individual investigative teams. The major challenges facing consortia such as ICVAS are to (1) obtain funding for establishment of a patient database and continued operations and (2) develop the collaborative culture and interactions to expand and take advantage of these economies of scale. Large patient databases for research on idiopathic scoliosis have also been described by Nancy Hadley Miller (Chapter 8) and by Carol Wise (Chapter 9) in this volume. Ongoing efforts described by both of these groups promise to identify additional loci that increase the risk of developing idiopathic scoliosis and its severity.

3. Building Large-Scale Patient Databases Useful for Scoliosis Research

The alternative to developing large patient databases focused on categories of disease (e.g., scoliosis and related spinal curves) is to develop the infrastructure to recruit patients with any disorder in a hospital or even national health provider system.

The most striking example of an effort to recruit all individuals in the national health-care system is being carried out by the deCode company to identify loci important for common genetic diseases. DeCode is based in the small country of Iceland, which has a population of approximately 320,000 that derives from an initial founder population (Price et al., 2009; http://www.decode.com). This population is subject to many of the same common diseases, such as diabetes, obesity, cancer, and cardiovascular disorders that afflict the population of the rest of the developed world. These disorders are also complex and involve the interactions between genetic background and environmental and life history factors. Disorders such as idiopathic scoliosis could also potentially be investigated in this population. Genetic mapping efforts have already been highly successful in identifying alleles that contribute to risk for major diseases. Even in 2009 alone, publications from or in collaboration with deCode have described loci that contribute risk for asthma, kidney stones, low bone mineral density, myocardial infarction, obesity, osteoarthritis, schizophrenia, and thyroid cancer (Evangelou et al. 2009, Gudbjartsson et al. 2009, Gudmundsson et al. 2009, Lindgren et al. 2009, Rujescu et al. 2009, Thorleifsson et al. 2009). Key to this success is the ability to integrate genotypic and medical records data from the Icelandic population, combined with patient databases from around the world.

The United States is a diverse nation with almost 1,000 times the population of Iceland and currently lacks a single electronic medical records and health-care finance structure. Thus, large-scale efforts at patient recruitment have taken place at the level of the hospital. One example of large-scale recruitment efforts within the United States has been at the Center for Applied Genomics at The

Children's Hospital of Philadelphia (CHOP, http://www.chop.edu). As with the deCode database efforts, patients with a diversity of disease, including diabetes, obesity, asthma, childhood cancers, and ADHD, are being recruited. This CHOP center plans to carry out high-throughput genotyping on samples from over 100,000 children in 3 years. One clear distinction from the deCode database is that this is predominantly a pediatric population, which may be particularly suitable for congenital disorders.

Both types of large-scale patient recruitment efforts hold great potential for scoliosis research. While the number of cases of rare disorders such as spondylocostal dysostosis may be small in any of these databases, these more comprehensive screens are more likely to identify larger populations of related complex disorders such as congenital scoliosis.

4. Future Use of Animal and Cell Culture Models for Scoliosis Research

Animal models are essential for scoliosis research, particularly in understanding the mechanisms leading to disease. Examples of the use of animal and cell culture models include the following:

- Validation of gene candidates identified in human genome-wide association studies,
- Identification of new genes that might play a role in relevant developmental processes, such as axial muscle development and innervation,
- Studies of gene–environment interactions focusing on environmental agents that contribute to scoliotic curves,
- Functional anatomical studies of the earliest steps in the development of spinal curvatures,
- Tests of potential therapies for scoliosis.

Animal models that display scoliosis or spinal curvatures can be generated by genetic or environmental disruptions. In the mouse, genetic models can arise spontaneously or result from phenotypes observed from gene targeting or transgenic efforts. Congenital defects of the spine are most readily visible in the tail, which makes up almost half of the vertebrae in an animal model such as the mouse. Progressive idiopathic scoliosis or kyphosis may be less readily detected, either from spontaneous or targeted mutations. To date, large-scale efforts to generate genetic models of scoliosis have not been carried out, but mutants have been isolated in the zebrafish and mouse from more general screens. Examples include an international phenotype-driven screen for novel developmental defects in mice generated by the powerful chemical mutagen N-ethyl-N-nitrosourea (Collins et al. 2007). Targeted mutation technology in the mouse has been useful in studies of somitogenesis that have advanced our understanding of congenital forms of scoliosis (Chapter 1). However, targeted mutations are of limited value currently for idiopathic scoliosis since we are only just beginning to understand the developmental origins, with only a single major locus identified to date (*CHD7*, Chapter 8).

In the zebrafish, the use of morpholino technology to generate "morphants" allows for the phenotypic effect of the loss of gene function to be evaluated. Somitogenesis has been studied in the zebrafish using morphants (reviewed in Wardle and Papaioannou 2008), which complement the array of mutations that have been generated for this model system. In addition to the zebrafish, other teleost models have proven useful for scoliosis research. Idiopathic scoliosis has also been analyzed in the guppy by genetic analysis seeking to identify genes regulating severity of spinal curves (Gorman et al. 2007).

Advances in high-throughput analysis of the transcriptome through microarray technology and of the proteome have been used to study developmental disorders. These approaches have been used to identify cycling genes in mouse somitogenesis (Dequéant et al. 2006), to identify somitogenesis genes disrupted in *Dll1* null embryos (Machka et al. 2005), *Hes7* null embryos (Niwa et al. 2007), *Dll3* and *Notch1* null embryos (Loomes et al. 2007, Sewell et al. 2009), and to identify oscillatory expression genes in human and mouse cell culture models (William et al. 2007). Human cell culture models are essential since the affected tissues cannot be sampled from human embryos. For idiopathic scoliosis, transcriptome and proteome analysis is hindered by our lack of understanding of the earliest mechanisms leading to spinal curvatures. If we are not certain what tissues and time points to focus on, high-throughput analysis of RNA or protein levels would likely only identify downstream effects of the original gene expression changes leading to scoliosis.

5. Utility of Systems Biology and Bioinformatic Approaches to Scoliosis Research

Bioinformatics will play an essential role in future scoliosis research. Bioinformatics will be essential to organize the radiographic images, clinical genetic and orthopedic medical records, pedigrees, outcomes data, genotypes, and genome sequence and also to allow for efficient searches of this data. Bioinformatics will also be required for developing the analytical tools to screen through this data for scoliosis research. While many "off-the-shelf" technologies can be applied, particularly for database efforts, the questions that are asked in scoliosis are unique. Curve progression is an essential question in the natural history of idiopathic scoliosis, and the outcomes data of surgical interventions are also relevant.

Systems biology, which combines bioinformatics, engineering, and biological approaches, will also play a role in basic research studies in animal models of scoliosis. One example of the applications of systems biology to scoliosis research is in our efforts to understand somitogenesis. When the genetic regulatory system of somitogenesis exceeds a small number of interactions, intuitive guesses alone are not sufficient to deduce regulatory pathways and generate clear hypotheses about somitogenesis. The identification of increasing numbers of Notch, Wnt, FGF, and other pathways with oscillatory gene expression clearly exceeds the capacity of human intuition. In such a situation, mathematical models can be created with component equations describing interactions of individual molecules in the system. These systems biological models are playing an increasing role in our understanding of somitogenesis. Most models have been based on delayed negative feedback loops of *Hes* genes, regulated by half-life of this transcription factor (Hirata et al. 2004, Lewis 2003, Monk 2003). The goal of these systems biological models is to be able to make predictions about relevant questions, e.g., how will a particular allele's decreased gene function disrupt somitogenesis, and then test this model through *in vitro* experiments. This will allow for the model to be refined through rounds of predictions and testing.

Summary

Developmental and genetic studies of the spine, genetic linkage to vertebral anomalies, and family-based association studies have led to advances in understanding the genetic causes of idiopathic and congenital scoliosis. Chapters in this volume have been prepared by research leaders who have been working to identify the genetic and developmental causes of idiopathic and congenital scoliosis. Technological advances in high-throughput sequencing, genotyping, bioinformatics, and medical imaging continue to push forward the limit of possible advances in scoliosis research. Combined with changes in the scale of human genetics research, we anticipate that the next decade could see an exponential increase in the number of genes associated with congenital and idiopathic scoliosis, and a greater understanding of the developmental mechanisms underlying "idiopathic" scoliosis.

References

Collins, F.S., Finnell, R.H., Rossant, J., and Wurst, W. 2007. A new partner for the International knockout mouse consortium. Cell 129(2):235.

Dequéant, M.L., Glynn, E., Gaudenz, K., Wahl, M., Chen, J., Mushegian, A., and Pourquié, O. 2006. A complex oscillating network of signaling genes underlies the mouse segmentation clock. Science 314:1595–1598.

Evangelou, E., Chapman, K., Meulenbelt, I., Karassa, F.B., Loughlin, J., Carr, A., Doherty, M., Doherty, S., Gómez-Reino, J.J., Gonzalez, A. et al. 2009. Large-scale analysis of association between GDF5 and FRZB variants and osteoarthritis of the hip, knee, and hand. Arthritis Rheum. 60(6):1710–1721.

Gorman, K.F., Tredwell, S.J., Breden, F. 2007. The mutant guppy syndrome curveback as a model for human heritable spinal curvature. Spine 32(7):735–741.

Gudbjartsson, D.F., Bjornsdottir, U.S., Halapi, E., Helgadottir, A., Sulem, P., Jonsdottir, G.M., Thorleifsson, G., Helgadottir, H., Steinthorsdottir, V., Stefansson, H. et al. 2009. Sequence variants affecting eosinophil numbers associate with asthma and myocardial infarction. Nat Genet. 41(3):342–347.

Gudmundsson, J., Sulem, P., Gudbjartsson, D.F., Jonasson, J.G., Sigurdsson, A., Bergthorsson, J.T., He, H., Blondal, T., Geller, F., Jakobsdottir, M. et al. 2009. Common variants on 9q22.33 and 14q13.3 predispose to thyroid cancer in European populations. Nat. Genet. 41(4):460–464.

Hirata, H., Bessho, Y., Kokubu, H., Masamizu, Y., Yamada, S., Lewis, J., and Kageyama, R. 2004. Instability of Hes7 protein is crucial for the somite segmentation clock. Nat. Genet. 36: 750–754.

International Human Genome Sequencing Consortium. 2004. Finishing the euchromatic sequence of the human genome. Nature 431(7011):931–945.

Lewis, J. 2003. Autoinhibition with transcriptional delay: a simple mechanism for the zebrafish somitogenesis oscillator. Curr. Biol. 13(16):1398–1408.

Lindgren, C.M., Heid, I.M., Randall, J.C., Lamina, C., Steinthorsdottir, V., Qi, L., Speliotes, E.K., Thorleifsson, G., Willer, C.J., Herrera, B.M. et al. 2009. Genome-wide association scan meta-analysis identifies three Loci influencing adiposity and fat distribution. PLoS Genet. 5(6):e1000508.

Loomes, K.M., Stevens, S.A., O'Brien, M.L., Gonzalez, D.M., Ryan, M.J., Segalov, M., Dormans, N.J., Mimoto, M.S., Gibson, J.D., Sewell, W., Schaffer, A.A., Nah, H.D., Rappaport, E.F., Pratt, S.C., Dunwoodie, S.L., and Kusumi, K. 2007. Dll3 and Notch1 genetic interactions model axial segmental and craniofacial malformations of human birth defects. Dev. Dyn. 236:2943–2951.

Machka, C., Kersten, M., Zobawa, M., Harder, A., Horsch, M., Halder, T., Lottspeich, F., Hrab de Angelis, M., and Beckers, J. 2005. Identification of Dll1 (Delta1) target genes during mouse embryogenesis using differential expression profiling. Gene Expr. Patterns 6:94–101.

Monk, N.A.M. 2003. Oscillatory expression of Hes1, p53 and NF-kB driven by transcriptional time delays. Curr. Biol. 13:1409–1413.

Niwa, Y., Masamizu, Y., Liu, T., Nakayama, R., Deng, C.X., and Kageyama, R. 2007. The initiation and propagation of Hes7 oscillation are cooperatively regulated by Fgf and notch signaling in the somite segmentation clock. Dev. Cell. 13:298–304.

Price, A.L., Helgason, A., Palsson, S., Stefansson, H., St Clair, D., Andreassen, O.A., Reich, D., Kong, A., and Stefansson, K. 2009. The impact of divergence time on the nature of population structure: an example from Iceland. PLoS Genet. 5(6):e1000505.

Rujescu, D., Ingason, A., Cichon, S., Pietiläinen, O.P., Barnes, M.R., Toulopoulou, T., Picchioni, M., Vassos, E., Ettinger, U., Bramon, E. et al. 2009. Disruption of the neurexin 1 gene is associated with schizophrenia. Hum. Mol. Genet. 18(5):988–996.

Thorleifsson, G., Holm, H., Edvardsson, V., Walters, G.B., Styrkarsdottir, U., Gudbjartsson, D.F., Sulem, P., Halldorsson, B.V., de Vegt, F., d'Ancona, F.C. et al. 2009. Sequence variants in the CLDN14 gene associate with kidney stones and bone mineral density. Nat. Genet. Jun 28 Epub ahead of print.

Sewell, W., Sparrow, D., Gonzalez, D.M., Smith, A., Eckalbar, W., Gibson, J., Dunwoodie, S.L., and Kusumi, K. 2009. Cyclical expression of the Notch/Wnt regulator Nrarp requires Dll3 function in somitogenesis. Dev. Biol. 329:400–409.

Turner, D.J., Keane, T.M., Sudbery, I., and Adams, D.J. 2009. Next-generation sequencing of vertebrate experimental organisms. Mamm. Genome. 20:327–338.

Wardle, F.C. and Papaioannou, V.E. 2008. Teasing out T-box targets in early mesoderm. Curr. Opin. Genet. Dev. 18:418–425.

William, D.A., Saitta, B., Gibson, J.D., Traas, J., Markov, V., Gonzalez, D.M., Sewell, W., Anderson, D.M., Pratt, S.C., Rappaport, E.F., and Kusumi, K. 2007. Identification of oscillatory genes in somitogenesis from functional genomic analysis of a human mesenchymal stem cell model. Dev. Biol. 305:172–186.

Index

Note: The letter 'f' and 't' following the locators refers to figures and tables respectively

K. Kusumi, S.L. Dunwoodie (eds.), *The Genetics and Development of Scoliosis*,
DOI 10.1007/978-1-4419-1406-4, © Springer Science+Business Media, LLC 2010

LaVergne, TN USA
18 February 2010
173498LV00001B/18/P

9 781441 914057